Gnamien trazie jean-marie

Maximize and transform your farming culture

Gnamien trazie jean-marie

Maximize and transform your farming culture

Practical agricultural guide

ScienciaScripts

Imprint
Any brand names and product names mentioned in this book are subject to trademark, brand or patent protection and are trademarks or registered trademarks of their respective holders. The use of brand names, product names, common names, trade names, product descriptions etc. even without a particular marking in this work is in no way to be construed to mean that such names may be regarded as unrestricted in respect of trademark and brand protection legislation and could thus be used by anyone.

Cover image: www.ingimage.com

This book is a translation from the original published under ISBN 978-620-6-71688-4.

Publisher:
Sciencia Scripts
is a trademark of
Dodo Books Indian Ocean Ltd. and OmniScriptum S.R.L publishing group

120 High Road, East Finchley, London, N2 9ED, United Kingdom
Str. Armeneasca 28/1, office 1, Chisinau MD-2012, Republic of Moldova, Europe
Printed at: see last page
ISBN: 978-620-7-91834-8

A- Food processing

Chapter 1: Introduction to food processing

Overview of the importance of food processing in adding value to agricultural products. Economic and nutritional benefits.

Chapter 2: Processing into Dairy Products

Explanation of the processing of dairy products such as milk, cheese, yoghurt and butter. Techniques and equipment required.

Chapter 3: Cereal processing

Methods for processing cereals into flour, bread, pasta and other by-products. Importance of grain quality and milling processes.

Chapter 4: Sugar crop processing

The process of transforming sugar cane and sugar beet into sugar, molasses and other sweet products. Impact on the local economy.

Chapter 5: Spice and herb processing

Techniques for drying, grinding and packaging spices and herbs. Added value and market potential.

Chapter 6: Oilseed processing

Processes for transforming oilseeds such as soy, sunflower and rapeseed into vegetable oils, margarine and other by-products.

Chapter 7: Brewing and winemaking

Brewing techniques for beer production and vinification techniques for wine production. Equipment requirements and key process steps.

B- Transformation into Chemical Products

Chapter 8: Introduction to Chemical Processing

Presentation of the different possibilities for chemical processing of agricultural products. The importance of innovation and sustainability.

Chapter 9: Conversion to biofuels

The process of producing biofuels from agricultural materials. Environmental impact and economic prospects.

Chapter 10: Transformation into Cosmetic Products

Techniques for transforming agricultural materials into cosmetics such as creams, lotions and essential oils. Market and regulations.

Chapter 11: Transformation into Pharmaceutical Products

Transformation of agricultural products into active pharmaceutical ingredients. Importance of quality and production standards.

Chapter 12: Conversion to Biodiesel and Bioethanol

Processes for producing biodiesel and bioethanol from specific crops. Applications and environmental benefits.

Chapter 13: The Starch Transformation

Methods for transforming starch into various industrial products. Applications in the food and non-food sectors.

Chapter 14: Enzymatic Hydrolysis Transformation

Using enzymes to transform agricultural materials into chemical products. Applications and benefits of enzyme technology.

Chapter 15: Transformation into polymerization

Polymerization process of agricultural materials to produce bioplastics and other materials. Innovations and applications.

Chapter 16: Transformation into Trans-Esterification

Transesterification techniques for the production of biodiesel and other products. Chemical processes and equipment required.

Chapter 17: Transformation into Green Chemistry

Principles of green chemistry applied to the processing of agricultural products. Sustainability and respect for the environment.

Chapter 18: Fiber conversion

Production of natural fibers from agricultural materials for textile and industrial applications. Techniques and market prospects.

C- Different types of agriculture

Chapter 19: Traditional agriculture

Description of traditional farming methods. Cultural importance and economic implications.

Chapter 20: Intensive agriculture

Intensive farming practices. Advantages and disadvantages in terms of productivity and environmental impact.

Chapter 21: Organic farming

Techniques and principles of organic farming. Certification and the organic market.

Chapter 22: Urban agriculture

Summary of the main points discussed. Reflections on the future of agricultural processing and types of agriculture for a sustainable future.

Sustainable, profitable production.

Introduction

Agriculture has always been at the heart of human survival and development. Since the earliest civilizations, man has constantly sought to improve his methods of growing and processing agricultural produce to meet his food and economic needs. It is against this backdrop that I have undertaken to write this practical guide, entitled "Maximizing and Transforming Your Farming: The Different Types of Agricultural Transformations". My main aim is to provide farmers, researchers, students and agricultural enthusiasts with a comprehensive and practical tool for maximizing the value of their agricultural produce.

The writing of this book stems from several observations and deep-seated motivations. As a farmer with a passion for sustainable development, I have observed the growing challenges facing agricultural producers: fluctuating commodity prices, climatic hazards, demographic pressure and increasing demands in terms of product quality and sustainability. Faced with these challenges, it is crucial to find innovative and effective ways of maximizing agricultural production while guaranteeing the sustainability and profitability of farms. The processing of agricultural products has proved to be a viable response to these challenges, not only extending the shelf life of products, but also diversifying farmers' sources of income.

This book is for anyone who wants to transform their approach to farming. It's not just about growing more, it's about growing better, and making the most of every product harvested. Agricultural processing, whether food or chemical, offers a multitude of possibilities for adding value to commodities. This practical guide will take you on a journey of discovery, exploring the different methods of processing agricultural products. You'll find detailed explanations of food transformation processes, such as the production of dairy products, the processing of cereals and sugar crops, and the processing of spices, herbs and oilseeds. Each section is designed to be both informative and accessible, with clear descriptions of the techniques and equipment required.

In addition, the book discusses chemical transformations of agricultural products, a fast-growing field that offers incredible opportunities for innovation and sustainability. You'll discover how agricultural materials can be converted into biofuels, cosmetics, pharmaceuticals and much more. The focus is on environmentally-friendly processes such as green chemistry and polymerization, which play a crucial role in the development of sustainable agriculture.

But this book doesn't just show you how to process your farm produce. It also explores the different types of agriculture, from traditional to intensive, organic and precision farming. Each type of farming is presented with its advantages, challenges and prospects, giving you a comprehensive overview of modern and innovative farming practices. Whether you're interested in urban agriculture, agroforestry or greenhouse farming, this book will provide you with the knowledge you need to maximize your production while respecting the environment.

In writing this book, I wanted to offer readers a work that is not only rich in information, but also inspiring. I want every reader to feel the urge and urgency to transform their farming practice, to see the infinite possibilities that processing agricultural products can offer. Each chapter is designed to capture your attention and give you the tools you need to put the concepts presented into practice. My writing style is dynamic and engaging, with the aim of making reading both enjoyable and rewarding.

Imagine being able to transform a simple corn crop into a diverse range of by-products: flour, oil, biofuels, even bioplastics. Think of the possibilities of adding value to your dairy products, creating high-quality cheeses and yogurts, or developing a line of cosmetics based on essential oils from your own crops. The economic benefits are obvious, but beyond that, there's the personal satisfaction of contributing to a more sustainable and innovative agriculture.

The future of agriculture lies in our ability to innovate and transform. This book is your guide to navigating this ever-changing landscape, providing you with the knowledge and tools you need to succeed. Whether you're a farmer looking to diversify your income, an agricultural student eager to learn the latest processing techniques, or simply passionate about sustainable development, this practical guide is for you.

Agricultural transformation is not just an economic necessity, it's also an opportunity to reinvent our relationship with the land and its resources. By maximizing and transforming our agricultural culture, we can create a future where every agricultural producer has the opportunity to prosper, where natural resources are used more efficiently, and where communities benefit from healthy, sustainable food.

Take this book as an invitation to explore new avenues, to push back the boundaries of what's possible, and to take an active part in the agricultural revolution of the 21st century. Together, we can transform not only our crops, but our world.

Happy reading and happy transformation!

Chapter 1: Introduction to Food Processing

Food processing is a crucial stage in the agricultural value chain. It plays an essential role in adding value to agricultural products, extending their shelf-life, improving their quality and diversifying income sources for farmers. This chapter presents an overview of the importance of food processing, highlighting its economic and nutritional benefits.

The Importance of Food Processing

Food processing is the process of modifying raw agricultural produce to create ready-to-eat food products or intermediate ingredients. This transformation can include processes such as drying, cooking, fermenting, grinding and many others. But why is it so important?

Increased added value: Processing agricultural products can add value to raw materials that are often under-valued. For example, transforming milk into cheese or yoghurt can considerably increase its market value. Similarly, processing fruit into jams or juices can produce higher value-added products than raw fruit.

Shelf-life extension: Agricultural products are often perishable. Food processing can extend the shelf life of products by making them less susceptible to spoilage. Drying fruit, for example, can extend shelf life by several months. Similarly, canning vegetables can preserve them for years.

Product diversification: Food processing offers the opportunity to diversify the products available on the market. This can include the creation of food products adapted to different tastes and cultural preferences. For example, wheat can be transformed into flour, then into bread, pasta, cookies and much more.

Reducing Post-Harvest Losses: A large proportion of agricultural produce is lost after harvest due to spoilage. Processing reduces these losses by stabilizing the product. For example, processing tomatoes into sauce or concentrate reduces losses due to rotting of fresh tomatoes.

Creating jobs and stimulating the local economy: Food processing requires specialized infrastructure and manpower, which can create jobs and stimulate the local economy. Small processing units can be set up in rural areas, creating employment opportunities for local communities.

Economic benefits

Food processing offers many economic benefits, not only for farmers, but also for local communities and the national economy.

Increased income for farmers: By processing their agricultural products, farmers can increase their income. The sale of higher value-added processed products generates additional income. For example, a farmer who processes his fruit into jam can sell these products at a much higher price than fresh fruit.

Price stabilization: Food processing helps stabilize the price of agricultural products by reducing dependence on seasonal fluctuations. By processing and storing produce, farmers can sell their processed products throughout the year, even in periods of low production.

Creating local and international markets: Processed products can be sold on local and international markets, opening up new business opportunities for farmers. For example, spices and dried herbs can be exported to international markets, creating new sources of income.

Encouraging innovation and entrepreneurship: Food processing encourages innovation and entrepreneurship by enabling farmers and small businesses to develop new products and markets. This can include the creation of organic, gluten-free or nutrient-enriched food products.

Reducing dependence on imports: By processing agricultural products locally, countries can reduce their dependence on imports of processed food products. This keeps value added in the local economy and reduces the trade deficit.

Nutritional benefits

Food processing can also have a positive impact on consumer nutrition and health. Here are just a few examples:

Improving food safety: Food processing extends the shelf life of products, thus contributing to food safety. For example, vegetable canning ensures that nutritious vegetables are available all year round, even in times of shortage.

Enriching food products: Processing enables food products to be enriched with vitamins, minerals and other essential nutrients. For example, milk can be enriched with vitamin D and calcium, which are beneficial for bone health.

Creating Functional Food Products: Functional food products are products that offer health benefits beyond their basic nutritional value. For example, probiotic yoghurts contain bacteria beneficial to digestive health. Processing creates these functional products.

Reducing food waste: Processing reduces food waste by using plant or animal parts that would not otherwise be consumed. For example, fruit skins can be used to produce jams or juices, thus reducing waste.

Food accessibility: Food processing makes food more accessible by making it easier to transport and store. For example, frozen vegetables can be stored for months, giving consumers access to nutritious vegetables all year round.

Tips for success in food processing

To succeed in food processing, it's important to follow certain best practices and focus on key aspects. Here are a few tips to maximize your chances of success:

Invest in Training and Education: It's essential to train and educate yourself in the various food processing techniques. Attend workshops, seminars and training courses to acquire the necessary skills.

Use Quality Equipment: Invest in quality processing equipment. Reliable and efficient equipment is essential to guarantee the quality of processed products and to increase production efficiency.

Comply with food safety standards: Make sure you comply with all food safety standards when processing products. Food safety is crucial to protecting consumer health and maintaining customer confidence.

Adopt Sustainable Practices: Use sustainable processing practices that minimize environmental impact. For example, use renewable energy sources and recycle production waste.

Market research: Before launching a new processed product, carry out market research to understand consumer needs and preferences. This will help you develop products that meet market expectations.

Constantly innovate: Be open to innovation and constantly seek to improve your products and transformation processes. Innovation is essential to staying competitive and responding to market trends.

Collaborate with Experts: Work with food processing experts, nutritionists and marketing specialists to improve the quality of your products and market them effectively.

Focus on Quality: Quality must be a top priority in food processing. High-quality products are essential for building customer loyalty and a good reputation.

Product diversification: Diversify your product range to cater for different market segments. Offer products for different consumption occasions and for different consumer groups.

Use Local Ingredients: Use local ingredients for food processing. This supports the local economy and reduces transportation costs and greenhouse gas emissions.

Successful Case Studies

To illustrate the importance and benefits of food processing, let's look at a few notable success stories:

Kwahu Dairy Cooperative, Ghana: This cooperative has invested in processing equipment to produce cheese and yoghurt from its members' milk. Thanks to these value-added products, farmers' incomes have risen by 40%, and the cooperative has been able to create local jobs.

Usine de Jus de Fruits de Koforidua, Côte d'Ivoire: This plant began by processing mangoes and pineapples into fruit juices and concentrates. Using modern pasteurization and packaging techniques, the plant was able to penetrate international markets and export its products to Europe and North America, generating substantial income for local producers.

La Fabrique de Céréales d'Addis-Abeba, Ethiopia: This factory processes teff, a local cereal, into flour and pasta. By adding value to this nutritious cereal, the company has not only contributed to local food security, but has also succeeded in positioning its products in niche international markets, attracting health-conscious consumers.

Colombo Spice Project, Sri Lanka: This spice processing project has enabled farmers to dry, grind and package their spices professionally. By improving the quality and presentation of the spices, the project has succeeded in accessing high-end markets and boosting producers' incomes.

These examples show how food processing can be an effective way of improving farm incomes, creating jobs, and promoting local economic development. By adopting good practices and taking advantage of market opportunities, farmers can process their produce profitably and sustainably.

Food processing is much more than a simple industrial process; it's an opportunity to create added value, diversify sources of income, and contribute to food security and nutrition. By investing in quality equipment, complying with food safety standards, and adopting sustainable practices, farmers can process their produce efficiently and profitably. Successful case studies show that, with the right strategies and a commitment to quality and innovation, food processing can be a pathway to a more prosperous and sustainable agricultural future.

Chapter 2: Processing into Dairy Products

Dairy processing is one of the oldest and most diversified branches of the food industry. It not only extends the shelf life of milk, but also creates an impressive variety of products, each with its own nutritional and economic benefits. This chapter explores in detail the processing of dairy products such as milk, cheese, yoghurt and butter, providing clear explanations of the techniques and equipment involved.

The Importance of Dairy Product Processing

Dairy product processing offers a number of advantages. It enables us to diversify our product offering, improve the profitability of dairy farms and meet the nutritional needs of different populations. It also plays a crucial role in food safety by stabilizing and extending the shelf life of dairy products.

Value added: Processing milk into higher value-added products, such as cheese and yoghurt, enables farmers to maximize their income. For example, the price of a liter of milk is much lower than that of a kilogram of cheese or a dozen pots of yogurt.

Extended shelf life: Fresh milk is a perishable product that needs to be consumed quickly. By processing it into products such as cheese, butter or milk powder, its shelf life can be extended by weeks, months or even years.

Product diversification: Processing enables us to create a variety of dairy products to suit different tastes and nutritional needs. For example, some consumers prefer plain yogurt, while others opt for flavored or probiotic-enriched yogurts.

Improved nutrition: Processed dairy products can be fortified with vitamins and minerals, improving their nutritional value. For example, milk can be enriched with vitamin D and calcium, which are beneficial for bone health.

Dairy Product Transformation Process

Dairy product processing involves several key stages, each requiring specific techniques and adapted equipment. Here's an overview of the main dairy products processed and the associated transformation processes:

Converting milk into cheese

Cheese is one of the oldest and most diverse of dairy products. Its production involves several stages, each playing a crucial role in the development of the final product's texture, taste and shelf life.

Pasteurization: The first step in the cheese-making process is to pasteurize the milk to eliminate pathogenic bacteria. The milk is heated to a precise temperature for a set period, then rapidly cooled.

Curdling: Once pasteurized, milk is coagulated using rennet or specific bacterial cultures. This process transforms the milk into a solid mass called curd.

Draining: The curds are then cut into small pieces to help remove the whey. This stage is crucial in determining the texture of the cheese. Longer draining will produce a firmer cheese.

Moulding and pressing: The drained curds are placed in moulds and pressed to remove the remaining whey. This stage gives the cheese its final shape.

Salting: Cheese is salted to improve its taste and preservation. Salting can be done on the surface or by immersion in brine.

Maturing: The cheese is then matured for a period that varies according to the type of cheese. Ripening develops the cheese's characteristic aromas and texture.

Processing Milk into Yogurt

Yogurt is a fermented dairy product appreciated for its creamy texture and probiotic benefits. Production is relatively simple and can be carried out on a small scale.

Pasteurization: As with cheese, milk for yogurt production is pasteurized to eliminate unwanted bacteria.

Inoculation: Once pasteurized and cooled, milk is inoculated with specific bacterial cultures, mainly Lactobacillus bulgaricus and Streptococcus thermophilus. These bacteria ferment lactose, producing lactic acid.

Incubation: The seeded milk is kept at a constant temperature (usually around 43°C) for several hours. This stage enables the bacteria to ferment the milk, thickening the product and developing the characteristic yogurt taste.

Cooling and packaging: Once fermentation is complete, the yogurt is rapidly cooled to stop fermentation and stabilize the product. It is then packaged in jars or bottles.

Converting milk into butter

Butter is a dairy product obtained by separating the fat from milk. Its production involves several processing stages.

Skimming: Milk is skimmed to separate the cream from the whey. This operation can be carried out using centrifuges or by gravity.

Pasteurization: The cream is pasteurized to eliminate pathogenic bacteria. This step is crucial to guarantee the safety of the final product.

Churning: Pasteurized cream is churned to separate the fat from the buttermilk. This operation transforms the liquid cream into a solid mass called butter.

Washing and kneading: The butter is washed to remove buttermilk residues and kneaded to obtain a homogeneous texture. Kneading also allows salt to be incorporated, if necessary.

Packaging: The butter is then packaged in blocks, trays or rolls, and refrigerated to extend its shelf life.

Techniques and equipment required

Dairy product processing requires specific techniques and equipment adapted to each type of product. Here is a list of equipment commonly used in dairy processing:

Pasteurizers: This equipment is used to heat milk to precise temperatures in order to eliminate pathogenic bacteria. They are available in various models, from plate pasteurizers to tubular pasteurizers.

Fermentation tanks: Used for yogurt production, these tanks maintain the milk at a constant temperature during the fermentation process. They are generally equipped with temperature and agitation control systems.

Churns: Churns are used for butter production. They separate the fat from the buttermilk by agitation. Modern churns can be programmed to control churning speed and duration.

Molds and presses: Used in cheese production, molds give the cheese its final shape, while presses remove the remaining whey. Presses can be manual or automatic.

Incubators: These are used to maintain bacterial cultures at optimum temperature during yogurt fermentation. They are often equipped with temperature and humidity control systems.

Cream separators: Used for skimming milk, these machines separate cream from whey by centrifugation. They are available in a range of models, from small manual separators to large industrial centrifuges.

Packaging equipment: This equipment is used to package processed dairy products. They include filling machines, labeling machines and packaging machines.

Practical Tips for Successful Dairy Processing

To succeed in dairy processing, it's important to follow certain best practices and focus on key aspects. Here are a few tips to maximize your chances of success:

Invest in Training and Education: It's essential to train and educate yourself in the various dairy processing techniques. Attend workshops, seminars and training courses to acquire the necessary skills.

Use Quality Ingredients: The quality of processed products depends largely on the quality of the ingredients used. Use fresh, high-quality milk to obtain top-quality finished products.

Comply with food safety standards: Make sure you comply with all food safety standards when processing dairy products. Food safety is crucial to protecting consumer health and maintaining customer confidence.

Adopt Sustainable Practices: Use sustainable processing practices that minimize environmental impact. For example, recycle production waste and use renewable energy sources.

Market research: Before launching a new processed product, carry out market research to understand consumer needs and preferences. This will help you develop products that meet market expectations.

Constantly innovate: Be open to innovation and constantly seek to improve your products and transformation processes. Innovation is essential to remain competitive and to respond to market trends.

Collaborate with Experts: Work with dairy processing experts, nutritionists and marketing specialists to improve the quality of your products and market them effectively.

Focus on Quality: Quality must be a top priority in dairy processing. High-quality products are essential for building customer loyalty and a good reputation.

Product diversification: Diversify your product range to cater for different market segments. Offer products for different consumption occasions and for different consumer groups.

Use modern technologies: Invest in modern technologies to improve processing efficiency and quality. Modern equipment enables more precise and efficient processes.

Successful Case Studies

To illustrate the possibilities offered by dairy processing, let's look at a few notable success stories:

La Fromagerie Artisanale de Beaufort, France: This small cheese dairy has made a name for itself by producing high-quality cheeses from local cow's milk. By using traditional techniques and investing in modern equipment, the dairy has been able to increase production while maintaining exceptional quality. Today, its cheeses are sold all over Europe.

Reykjavik Yogurt Factory, Iceland: This factory began by producing skyr yogurt, an Icelandic specialty. By using specific bacterial cultures and optimizing fermentation processes, the plant succeeded in producing a protein-rich yogurt appreciated for its thick texture. Thanks to an effective marketing strategy, skyr is now exported to many countries and has become a flagship product of the Icelandic dairy industry.

La Coopérative Laitière de Fès, Morocco: This cooperative has invested in the processing of milk into butter and fromage frais. By using sustainable production methods and complying with food safety standards, the cooperative has been able to increase its members' incomes while supplying quality dairy products to local markets. The cooperative has also set up training programs for its members, improving their skills and ability to produce quality products.

The Dairy Processing Project in Pune, India: This project enabled rural women to join forces to process milk into yoghurt, cheese and butter. By providing processing equipment and training, the project has contributed to women's economic empowerment and improved local food security. Processed products are sold on local markets and in larger towns, generating additional income for rural families.

These examples show how dairy processing can be an effective way of improving farm incomes, creating jobs and promoting local economic development. By adopting good practices and taking advantage of market opportunities, farmers can process their dairy products profitably and sustainably.

Chapter 3: Cereal processing

Cereal processing is a fundamental pillar of the global food industry. Cereals, such as wheat, corn, rice, barley, rye and oats, are the basic ingredients for a multitude of food products. This chapter explores in detail how cereals are processed into flour, bread, pasta and other by-products. It also emphasizes the importance of grain quality and milling processes in obtaining high-quality end products.

The Importance of Cereals in the Diet

Cereals have played a central role in the human diet for thousands of years. They are an essential source of carbohydrates, fiber, vitamins and minerals. Their processing makes it possible to diversify the diet and meet the nutritional needs of different populations. Here's why cereal processing is so crucial:

Food diversity: Cereal processing produces a variety of food products, from flour to breads, pasta, cookies and breakfast cereals. This diversity helps enrich the daily diet and meet consumers' taste preferences.

Extended shelf life: Processed cereals, such as flour and pasta, have a longer shelf life than whole grains, making them easier to store and transport. This contributes to food security by ensuring continuous availability of basic food products.

Improved digestibility: Processing cereals improves their digestibility and bioavailability. For example, milling and cooking break down complex components, making nutrients more accessible to the body.

Creating added value: Grain processing adds value to raw grain, offering farmers and businesses additional income opportunities. For example, the sale of flour or by-products is generally more lucrative than that of unprocessed grain.

Cereal processing methods

The processing of cereals involves several stages, each of which is essential for obtaining high-quality products. Here are the main methods used to transform cereals into flour, bread, pasta and other by-products:

Cereal milling

Milling is the process of transforming cereal grains into flour. This process involves several stages, each of which plays a crucial role in the quality of the flour produced.

Cleaning and preparation: Before milling, cereal grains must be carefully cleaned to remove impurities such as stones, dust and plant residues. This step is crucial to guarantee the purity of the flour.

Milling: The cleaned grains are then milled using grinding mills or roller mills. Grinding reduces the grains to finer particles, facilitating their transformation into flour.

Sifting: After grinding, the flour is sifted to separate the different fractions. Sifting produces flours of different textures, such as white flour, wholemeal flour and semolina flour.

Enrichment: In some cases, flour can be enriched with vitamins and minerals to enhance its nutritional value. For example, wheat flour can be enriched with iron and folic acid.

The transformation of flour into bread

Bread is one of the most common products derived from flour. Bread-making involves several essential stages:

Preparing the dough: Flour is mixed with water, yeast, salt and other ingredients to form a dough. Yeast is a key ingredient in the fermentation process.

Kneading: The dough is kneaded to develop the gluten, a protein that gives bread its structure and elasticity. Kneading can be carried out manually or using mechanical mixers.

Fermentation: The kneaded dough is left to ferment for a set period. Fermentation allows the yeast to produce carbon dioxide, making the dough rise and improving the texture of the bread.

Shaping: After fermentation, the dough is shaped into loaves or baguettes. This stage gives the bread its final shape.

Baking: The shaped bread is then baked in an oven at specific temperatures. Baking fixes the structure of the bread and develops its crust and aromas.

Converting flour into pasta

Pasta is another popular by-product of flour. Pasta-making involves several stages:

Mixing the dough: Durum wheat semolina flour is mixed with water to form a dough. Unlike bread, no yeast is needed to make pasta.

Laminating and extrusion: Dough is laminated into sheets or extruded through dies to form different pasta shapes, such as spaghetti, macaroni and lasagne.

Drying: Fresh pasta is dried to extend its shelf life. Drying can be carried out in the open air or in industrial dryers.

Packaging: Dried pasta is packaged in packets or tins for sale. Packaging protects the pasta from moisture and contaminants.

Other Cereal Derivatives

In addition to flour, bread and pasta, cereals can be transformed into a multitude of other food products:

Breakfast cereals: Cereal grains can be processed into flakes, granules or extrudates to create breakfast cereals. These products can be enriched with vitamins and minerals and flavored to appeal to consumers.

Cookies and cakes: Cereal flour is also used to make cookies, cakes and pastries. These products require specific baking techniques and complementary ingredients such as sugar, eggs and leavening agents.

Snacks and crackers: Cereals can be transformed into snacks and crackers by extrusion, baking or frying. These products are often seasoned and flavored to meet consumer preferences.

Bakery products: In addition to bread, bakery products include brioches, croissants, muffins and other pastries. The transformation of flour into these products requires specific baking and fermentation techniques.

Importance of Grain Quality

The quality of cereal grains is essential to guarantee the quality of processed products. Several factors influence grain quality:

Cereal variety: The variety of cereal used influences the quality of the flour and derived products. For example, durum wheat is preferred for pasta production because of its high protein and gluten content.

Growing conditions: Growing conditions, such as climate, soil and farming practices, affect grain quality. Sustainable farming practices and proper crop management ensure high-quality grain.

Harvesting and storage: The way grain is harvested and stored influences its quality. Grain must be harvested at the right time and stored in optimal conditions to avoid deterioration and infestation.

Cleaning and preparation: Before processing, the grains must be carefully cleaned to remove impurities. This step is crucial to obtaining high-quality finished products.

Practical Tips for Successful Cereal Processing

To succeed in cereal processing, it's important to follow certain good practices and focus on key aspects. Here are a few tips to maximize your chances of success:

Invest in Quality Equipment: Use high-quality milling and processing equipment. Well-calibrated mills and efficient cooking machines are essential to guarantee the quality of processed products.

Use high-quality grains: The quality of cereal grains has a direct impact on the quality of finished products. Choose cereal varieties suited to the desired processing and ensure that the grains are clean and well preserved.

Comply with food safety standards: Make sure you comply with all food safety standards when processing cereals. Food safety is crucial to protecting consumer health and maintaining customer confidence.

Adopt Sustainable Practices: Use sustainable processing practices that minimize environmental impact. For example, recycle milling by-products and use renewable energy sources.

Market research: Before launching a new processed product, carry out market research to understand consumer needs and preferences. This will help you develop products that meet market expectations.

Constantly innovate: Be open to innovation and constantly seek to improve your products and transformation processes. Innovation is essential to staying competitive and responding to market trends.

Collaborate with Experts: Work with grain processing experts, nutritionists and marketing specialists to improve the quality of your products and market them effectively.

Focus on Quality: Quality must be a top priority in cereal processing. High-quality products are essential for building customer loyalty and a good reputation.

Product diversification: Diversify your product range to cater for different market segments. Offer products for different consumption occasions and for different consumer groups.

Use modern technologies: Invest in modern technologies to improve processing efficiency and quality. Modern equipment enables more precise and efficient processes.

Successful Case Studies

To illustrate the possibilities offered by cereal processing, let's look at a few notable success stories:

La Minoterie de Talavera, Spain: This family-run flour mill has combined tradition and modernity to produce high-quality flours. By using carefully selected local grains and investing in modern milling equipment, the mill has succeeded in distinguishing itself on the market.

It produces a diversified range of flours for artisanal and industrial bakeries, meeting the needs of both local bakers and major retail chains.

L'Usine de Pâtes de Parme, Italy: This pasta factory is famous for the quality of its products, made from high-quality durum wheat semolina. Using traditional production methods combined with cutting-edge technology, the factory has succeeded in producing pasta with authentic textures and tastes. Parma pasta is exported all over the world, a testament to the plant's reputation and excellence.

La Fabrique de Biscuits, Milton Keynes, UK: This innovative cookie factory uses ancient grains and natural ingredients to produce nutritious, tasty cookies. By working with local farmers to source high-quality grains, the factory has been able to meet growing consumer demand for healthy, ethical products. Milton Keynes cookies have become a popular choice in supermarkets and organic stores.

Kaduna Cereal Processing Project, Nigeria: This community-based project has transformed local cereals into flour, bread and snacks for local and regional markets. By providing processing equipment and training to farmers, the project has improved food security and income for rural families. Processed products are sold on local markets and in larger towns, generating additional income for rural families and contributing to the region's economic stability.

These examples show how grain processing can be an effective way of improving farm incomes, creating jobs and promoting local economic development. By adopting good practices and taking advantage of market opportunities, farmers can process their cereal products profitably and sustainably.

Chapter 4: Sugar crop processing

The processing of sugar crops, such as sugar cane and sugar beet, into sugar, molasses and other sweet products, is an industry of great economic and social importance. This chapter explores in depth the processes involved in transforming these sugar crops, the techniques and equipment required, and the economic and social impact on local communities.

Importance of sugar crops

Sugar crops play a vital role in the economies of many tropical and temperate countries. Sugar cane and sugar beet are the main sources of sugar, an essential ingredient in the human diet and a key component of many food and industrial products. Here are just a few reasons why sugar crops are crucial:

Source of income: Sugar cane and sugar beet production and processing generate significant income for local farmers and industries. Sugar is a widely consumed commodity, ensuring constant demand.

Job creation: The sugar industry creates jobs at various stages of the value chain, from the cultivation and harvesting of sugar plants to the processing and distribution of finished products.

Product diversification: Sugar crops can be processed into a variety of products, including sugar, molasses, syrup, ethanol, and even food and cosmetics, thus diversifying sources of income.

Economic impact: Sugar crop processing has a significant impact on the local economy, stimulating industrial development, infrastructure and trade in the producing regions.

Sugar Cane Transformation Process

Sugar cane is a tropical plant cultivated mainly for its high sucrose content. The transformation of sugar cane into sugar and other sweet products involves several complex stages, each requiring specific techniques and adapted equipment.

Sugar Cane Harvesting and Preparation

Harvesting: Sugar cane can be harvested either manually or mechanically. The cane stalks are cut at the base and transported to the processing plant.

Cleaning and preparation: The cane stalks are cleaned to remove impurities such as leaves, earth and debris. They are then cut into smaller pieces to facilitate juice extraction.

Juice extraction

Crushing: Cane pieces are crushed using cane mills to extract the sweet juice. This operation separates the juice from the fibrous residue known as bagasse.

Clarification: The extracted juice is clarified to remove impurities. Clarifying agents, such as lime or aluminum sulfate, are added to precipitate impurities, which are then filtered.

Evaporation and Crystallization

Evaporation: The clarified juice is concentrated by evaporation in vacuum evaporators. This step reduces the juice's water content, transforming it into a thick syrup.

Crystallization: The concentrated syrup is then crystallized in vacuum cookers. The sugar crystallizes out of the syrup, forming raw sugar crystals.

Separation and Refining

Centrifugation: The sugar crystals are separated from the residual syrup by centrifugation. The remaining syrup, called molasses, can be used to produce other products.

Refining: Raw sugar can be refined to produce refined white sugar. This stage includes washing, dissolving, filtering and recrystallizing the sugar to obtain a pure, white product.

Sugar beet transformation process

Sugar beet is a temperate crop grown for its high sucrose content. The process of transforming sugar beet into sugar is similar to that of sugar cane, but with a few specific differences.

Sugar beet harvesting and preparation

Harvesting: Beets are harvested mechanically, using machines that pull the roots out of the soil and clean them roughly.

Cleaning and preparation: Beets are washed to remove soil and debris, then cut into thin strips called cossettes to facilitate juice extraction.

Extraction and Diffusion

Diffusion: The cossettes are immersed in hot water in diffusers to extract the sweet juice by osmosis. This process is more efficient for beet than the crushing used for cane.

Clarification: The extracted juice is clarified to remove impurities in the same way as cane juice.

Evaporation and Crystallization

Evaporation: The clarified juice is concentrated by evaporation to form a thick syrup.

Crystallization: Concentrated syrup is crystallized to form raw sugar crystals, which are then separated by centrifugation.

Refining

Refining: Raw beet sugar is refined in a similar way to cane sugar, by washing, dissolving, filtering and recrystallizing.

Sugar crop by-products

In addition to sugar, the processing of sugar cane and sugar beet produces a variety of by-products and derivatives that are economically important.

Molasses

Molasses is the residual syrup left after sugar crystallization. It is rich in minerals and non-crystallizable sugar. Molasses can be used in a variety of ways:

Animal feed: Molasses is often used as a feed supplement for livestock because of its high energy content.

Fermentation: Molasses is an important raw material for the production of ethanol by fermentation, used as a biofuel or in the alcoholic beverage industry.

Food products: Molasses is used as an ingredient in the manufacture of certain foods and beverages, such as gingerbread, syrups and sauces.

Bagasse

Bagasse is the fibrous residue left after cane juice extraction. It has several industrial uses:

Biofuel: Bagasse is used as biofuel to produce energy in sugarcane processing plants. It can also be burned to generate steam and electricity.

Paper manufacture: Bagasse is also used as a raw material in the manufacture of paper and cardboard, offering a sustainable alternative to wood.

Building materials: Bagasse can be processed into building materials such as particleboard and insulation board.

Ethanol

Ethanol can be produced from the fermentation of molasses or cane juice. It has several important applications:

Biofuel: Ethanol is used as a biofuel for vehicles, reducing dependence on fossil fuels and greenhouse gas emissions.

Chemical industry: Ethanol is an important solvent in the chemical and pharmaceutical industries, used in the manufacture of various products.

Alcoholic beverages: Ethanol is also used in the production of alcoholic beverages such as rum and vodka.

Economic and social impact

Sugar crop processing has a significant impact on the economy and society of sugar-producing regions. Here are some of the main economic and social impacts:

Job creation: The sugar industry creates jobs at various stages of the value chain, from cultivation and harvesting to processing and distribution. This helps to reduce poverty and improve living conditions in local communities.

Industrial development: Processing sugar crops stimulates industrial development by encouraging the installation of processing plants, transport infrastructure and related services. This generates investment and opportunities for economic growth.

Income stability: Diversification of products derived from sugar crops enables farmers and companies to stabilize their income by reducing dependence on a single source of revenue. By-products such as molasses, bagasse and ethanol offer additional income opportunities.

Support for rural communities: The sugar industry plays a crucial role in supporting rural communities by providing jobs, infrastructure and essential services. It also contributes to food security by guaranteeing a constant supply of sugar products.

Environment and Sustainability: Sugar crop processing can have positive environmental impacts by promoting the use of by-products as renewable energy sources. However, it also requires responsible management to minimize negative impacts such as deforestation and water pollution.

Practical Tips for Successful Sugar Crop Processing

To succeed in sugar crop processing, it's important to follow certain good practices and focus on key aspects. Here are a few tips to maximize your chances of success:

Invest in modern equipment: Use high-quality, modern processing equipment. Efficient cane mills, well-calibrated evaporators and centrifuges are essential to guarantee the quality of processed products.

Use Quality Raw Materials: The quality of raw materials has a direct impact on the quality of finished products. Make sure that sugar cane and sugar beet are harvested at the right time and are free from impurities.

Comply with food safety standards: Make sure you comply with all food safety standards when processing sugar crops. Food safety is crucial to protecting consumer health and maintaining customer confidence.

Adopt Sustainable Practices: Use sustainable processing practices that minimize environmental impact. For example, recycle processing by-products and use renewable energy sources.

Market research: Before launching a new processed product, carry out market research to understand consumer needs and preferences. This will help you develop products that meet market expectations.

Constantly innovate: Be open to innovation and constantly seek to improve your products and transformation processes. Innovation is essential to staying competitive and responding to market trends.

Collaborate with Experts: Work with sugar crop processing experts, nutritionists and marketing specialists to improve the quality of your products and market them effectively.

Focus on Quality: Quality must be a top priority in sugar crop processing. High-quality products are essential for building customer loyalty and a good reputation.

Product diversification: Diversify your product range to cater for different market segments. Offer products for different consumption occasions and for different consumer groups.

Use modern technologies: Invest in modern technologies to improve processing efficiency and quality. Modern equipment enables more precise and efficient processes.

Successful Case Studies

To illustrate the possibilities offered by sugar crop processing, let's look at a few notable success stories:

Sucrerie de Réunion, France: This sugar mill has succeeded in setting itself apart by producing high-quality cane sugar for local and international markets. By investing in modern equipment and adopting sustainable practices, the sugar plant has succeeded in increasing production while reducing its environmental impact. It also produces ethanol from molasses, diversifying its sources of revenue.

Nebraska Beet Plant, USA: This sugar beet processing plant has implemented advanced processing techniques to produce high-quality refined sugar. By using state-of-the-art technologies for beet juice extraction and clarification, the plant has been able to improve production efficiency and reduce costs. It has also developed by-products such as sweet snacks and syrups, responding to growing consumer demand for innovative products.

Kwazulu-Natal Sugar Cane Processing Project, South Africa: This community-based project processes sugar cane into sugar and molasses for local and regional markets. By providing processing equipment and training to farmers, the project has improved food security and income for rural families. Processed products are sold on local markets and in larger towns, generating additional income for rural families and contributing to the region's economic stability.

These examples show how processing sugar crops can be an effective way of improving farm incomes, creating jobs and promoting local economic development. By adopting good practices and taking advantage of market opportunities, farmers can process their sugar products profitably and sustainably.

Chapter 5: Spice and herb processing

Spices and aromatic herbs occupy a central place in the world's cuisine, bringing flavor, aroma and color to dishes. Processing these plants adds value and extends their shelf life, while opening up potential markets for producers. This chapter explores drying, grinding and packaging techniques for spices and herbs, focusing on economic benefits and market opportunities.

The Importance of Spices and Herbs

Spices and herbs have a long history of use, not only for their culinary properties, but also for their medicinal and preservative benefits. Here's why they're essential:

Culinary diversity: Spices and herbs are used in a variety of cuisines to add unique and complex flavors to dishes. They are indispensable in the preparation of many traditional and modern recipes.

Medicinal properties: Many spices and herbs have medicinal properties. Turmeric, for example, is known for its anti-inflammatory properties, while mint is used for its soothing effects on the digestive system.

Food preservation: Some spices, such as cloves and cinnamon, have antimicrobial and antioxidant properties that help preserve foods and extend their shelf life.

Economic opportunities: Processing spices and herbs enables farmers to diversify their income and access new markets. Processed products, such as powders, essential oils and extracts, have a higher added value than raw products.

Spice and Herb Processing Techniques

The processing of spices and herbs involves several key stages: drying, grinding and packaging. Each of these stages is crucial in guaranteeing the quality and shelf life of the finished products.

Drying spices and herbs

Drying is the first stage in the processing of spices and herbs. It reduces the water content of the plants, preventing the growth of micro-organisms and thus extending their shelf life. Here are the main drying methods:

Air-drying: Air-drying is the oldest and simplest method. Spices and herbs are spread out on clean, dry surfaces, away from direct sunlight. This method is economical, but depends on climatic conditions.

Sun-drying: Sun-drying involves exposing plants to direct sunlight. Although quick, this method can result in a loss of color and flavor due to exposure to ultraviolet rays. It is best suited to arid climates.

Kiln drying : Oven drying allows precise control of temperature and humidity. Plants are placed in ovens at low temperatures (40-50°C) until they are completely dry. This method is faster and offers better quality control.

Dehydration drying: Electric dehydrators use a stream of hot air to remove moisture from plants. This efficient method preserves the color, flavor and nutrients of spices and herbs.

Spice and herb grinding

Grinding is the next stage after drying. It involves reducing the dried plants to a fine powder or smaller pieces, ready for culinary or industrial use. Grinding techniques vary according to production scale and desired texture.

Manual grinding: Manual grinding using mortars and pestles is traditionally used for small quantities. This method enables the texture of the final product to be controlled, but is laborious and unsuitable for large-scale production.

Mechanical milling: Electric or mechanical mills are used to grind larger quantities of spices and herbs. These machines can be adjusted to obtain different granulometries, from fine powder to coarse pieces.

Cryogenic grinding: This technique uses extremely low temperatures to grind spices and herbs. Cryogenic grinding preserves aromas and essential oils, but requires specialized equipment and high operating costs.

Spice and herb packaging

Packaging is the final stage of processing. It involves wrapping finished products to protect them from moisture, light and contaminants, while facilitating storage and transportation.

Sachet packaging: Plastic, aluminum or paper sachets are commonly used to package spices and herbs. They are airtight, lightweight and practical for storage.

Vacuum packing: Vacuum packing involves removing the air from the sachets before sealing them. This method extends shelf life by preventing oxidation and microbial growth.

Glass jars: Glass jars are often used for high-quality spices. They offer excellent protection against humidity and light, while being reusable and aesthetically pleasing.

Labeling: Labeling is a crucial step in the packaging process. Labels must include product information such as name, harvest date, expiry date, storage instructions and any certifications.

Added value and market potential

Spice and herb processing offers significant added value and opens up new markets for growers. Here are some key aspects of added value and market potential:

Added value: Processed spices and herbs, such as powders, spice blends, essential oils and extracts, have a higher added value than raw products. This added value translates into higher profit margins for producers.

Extended shelf life: Processing spices and herbs extends their shelf life, making them easier to store and transport. This enables producers to sell their products all year round, even out of season.

Access to international markets: Processed spices and herbs meet the quality and food safety standards demanded by international markets. This opens up export opportunities to countries where demand for these products is strong.

Product diversification: Growers can diversify their product range by offering spices and herbs in different forms (powders, dried leaves, essential oils, extracts) and for different uses (culinary, medicinal, cosmetic).

Responding to Consumer Demand: Consumer demand for natural, organic and high-quality products is constantly increasing. Spices and herbs processed in a sustainable and ethical way meet these expectations and can attract a premium market segment.

Practical Tips for Successful Spice and Herb Processing

To succeed in processing spices and herbs, it's important to follow certain good practices and focus on key aspects. Here are a few tips to maximize your chances of success:

Invest in Quality Equipment: Use high-quality drying, grinding and packaging equipment. Efficient, well-calibrated machinery is essential to guarantee the quality of processed products.

Use quality raw materials: The quality of spices and herbs has a direct impact on the quality of finished products. Make sure plants are harvested at the right time and are free from impurities.

Comply with food safety standards: Make sure you comply with all food safety standards when processing spices and herbs. Food safety is crucial to protecting consumer health and maintaining customer confidence.

Adopt Sustainable Practices: Use sustainable processing practices that minimize environmental impact. For example, recycle processing by-products and use renewable energy sources.

Market research: Before launching a new processed product, carry out market research to understand consumer needs and preferences. This will help you develop products that meet market expectations.

Constantly innovate: Be open to innovation and constantly seek to improve your products and transformation processes. Innovation is essential to staying competitive and responding to market trends.

Collaborate with Experts: Work with spice and herb processing experts, nutritionists and marketing specialists to improve the quality of your products and market them effectively.

Focus on Quality: Quality must be a top priority when processing spices and herbs. High-quality products are essential for building customer loyalty and a good reputation.

Product diversification: Diversify your product range to cater for different market segments. Offer products for different consumption occasions and for different consumer groups.

Use modern technologies: Invest in modern technologies to improve processing efficiency and quality. Modern equipment enables more precise and efficient processes.

Successful Case Studies

To illustrate the possibilities offered by spice and herb processing, let's look at a few notable success stories:

Le Projet de Transformation des Herbes Aromatiques de Provence, France: This community project has transformed local aromatic herbs such as thyme, rosemary and lavender into essential oils, dried herb sachets and spice blends. Using traditional drying and distillation methods combined with modern packaging techniques, the project has succeeded in creating a range of high-quality products for local and international markets.

Kerala Spice Grinding Plant, India: This plant has implemented advanced cryogenic grinding techniques to produce fine, aromatic spice powders such as turmeric, black pepper and cardamom. By investing in modern equipment and complying with food safety standards, the plant has been able to penetrate premium spice markets, responding to growing consumer demand for high-quality spices.

La Coopérative de Transformation d'Herbes de Marrakech, Morocco: This cooperative has enabled rural women to come together to process local herbs such as mint, coriander and parsley into dried products and essential oils. By providing processing equipment and training, the cooperative has improved food security and income for rural families. Processed products are sold on local markets and in larger towns, generating additional income for rural families and contributing to the region's economic stability.

These examples show how processing spices and herbs can be an effective way of improving farm incomes, creating jobs and promoting local economic development. By adopting good practices and taking advantage of market opportunities, farmers can process their spice products profitably and sustainably.

Chapter 6: Oilseed processing

The processing of oilseeds such as soy, sunflower and rapeseed is essential for the production of vegetable oils, margarine and other by-products. These products play a crucial role in human and animal nutrition, as well as in various industries. This chapter explores oilseed processing, the techniques and equipment required, and the economic importance of these transformations.

Importance of oilseeds

Oilseeds are grown primarily for their high oil and protein content. They are of great economic and nutritional importance for the following reasons:

Source of Vegetable Oils: Oils extracted from oilseeds are widely used in cooking, baking and the preparation of various processed foods. They are also used in the manufacture of margarine and shortenings.

Animal feed: Oilcake, the residue from oil extraction, is a valuable source of protein for animal feed. They are used as a supplement in cattle, poultry and fish rations.

Industrial applications: Vegetable oils and their derivatives are used in the manufacture of cosmetics, soaps, biofuels, lubricants and other industrial products.

Food security: Oilseeds contribute to food security by providing essential nutrients such as unsaturated fatty acids, vitamins and proteins. They play a key role in balanced diets and global nutrition.

Oilseed Transformation Processes

Processing oilseeds into vegetable oils and other by-products involves several key stages. Each stage requires specific techniques and adapted equipment to guarantee efficient extraction and optimum quality of the finished products.

Seed preparation

The first stage of processing involves preparing the oilseeds for oil extraction. This includes cleaning, hulling, crushing and packaging.

Cleaning: Oilseeds are cleaned to remove impurities such as stones, plant debris and dust. This step is crucial to guarantee oil purity and avoid wear and tear on processing equipment.

Dehulling: Some seeds, such as sunflower and soybean, need to be dehulled to remove hulls or husks. Dehulling increases oil yield and reduces the fiber content of oilcake.

Crushing: Shelled seeds are crushed into small particles to facilitate oil extraction. Crushing is carried out using hammer mills, roller mills or grinding mills.

Conditioning: The ground particles are conditioned by heating them to moderate temperatures. Conditioning releases the oils contained in the seed cells and prepares the material for further processing.

Oil extraction

Oil extraction is the central phase in oilseed processing. It can be carried out by mechanical pressure, solvent extraction or a combination of the two.

Mechanical pressure: Mechanical pressure is a traditional method of oil extraction. Packaged seed particles are subjected to high pressure in screw or hydraulic presses to extract the oil. This method is simple and requires no chemical solvents, but it generally leaves a residual quantity of oil in the cakes.

Solvent extraction : Solvent extraction is a more modern and efficient method. The seed particles are mixed with a solvent (usually hexane) which dissolves the oil. The mixture is then heated to evaporate the solvent, leaving behind the extracted oil. This method achieves a higher oil yield than mechanical pressing.

Combined extraction: Some plants use a combination of mechanical pressure and solvent extraction. This approach maximizes oil yield and optimizes process efficiency.

Oil refining

Crude oil extracted from oilseeds contains impurities such as phospholipids, pigments, free fatty acids and waxes. Oil refining is necessary to improve its quality and stability. The refining process comprises several stages:

Degumming: Degumming is the first step in refining. It consists of removing phospholipids and gums by treating the crude oil with hot water or acid. Impurities are then separated by centrifugation.

Neutralization: Neutralization, also known as deacidification, involves removing free fatty acids from the crude oil. The oil is treated with an alkaline solution (often caustic soda), which reacts with the free fatty acids to form soaps. These soaps are then removed by centrifugation.

Decolorization: Decolorization removes pigments and colorants from the oil. The oil is mixed with clays or activated carbons, which adsorb the pigments. The mixture is then filtered to obtain a clearer oil.

Deodorization: Deodorization is the final step in the refining process. It consists of eliminating the volatile compounds responsible for undesirable odors and tastes. The oil is heated in a vacuum, and the steam produced carries away the volatile compounds.

Transformation into Derivatives

In addition to vegetable oil, oilseeds can be processed into various by-products such as margarine, shortenings, emulsifiers and vegetable proteins.

Margarine: Margarine is made from refined vegetable oils, which are hydrogenated to partially solidify them. The hydrogenated oil is mixed with water, salt, emulsifiers, colorants and flavorings to form a homogeneous paste. The margarine is then cooled and packaged.

Shortenings: Shortenings are solid vegetable fats used in baking and cooking for their texture and cooking properties. They are made from partially hydrogenated vegetable oils, which are then cooled and kneaded to obtain the desired consistency.

Emulsifiers : Refined vegetable oils can be converted into emulsifiers, which are used to stabilize water-oil mixtures in a variety of food products, such as sauces, ice creams and baked goods. Lecithins and mono- and diglycerides are common examples of emulsifiers derived from oilseeds.

Plant proteins: Oilcake, the residue from oil extraction, is rich in protein and can be transformed into plant protein concentrates or isolates. These products are used in animal and human nutrition for their high protein content and functional properties.

Economic Importance of Oilseed Processing

Oilseed processing has a significant economic impact on several levels:

Increased added value: Processing oilseeds into vegetable oils and other by-products adds value to the raw materials. Processed products can be sold at higher prices than raw seeds, thus increasing the income of producers and processors.

Job creation: The oilseed processing industry creates jobs at various stages of the value chain, from the cultivation and harvesting of seeds to the processing, packaging and distribution of finished products. This helps reduce poverty and improve living conditions in rural areas.

Product diversification: Oilseed processing diversifies the products available on the market. Vegetable oils, margarine, vegetable proteins and emulsifiers are used in a variety of industries, increasing market opportunities and economic resilience.

Improved food security: Products derived from oilseeds, such as vegetable oils and proteins, contribute to food security by providing essential nutrients. Protein-rich oilcakes are an important source of feed for livestock, supporting the production of meat, milk and other animal products.

Industrial development: Oilseed processing stimulates industrial development by encouraging the installation of processing plants, transport infrastructure and related services. This generates investment and economic growth opportunities.

Practical Tips for Successful Oilseed Processing

To succeed in oilseed processing, it's important to follow certain good practices and focus on key aspects. Here are a few tips to maximize your chances of success:

Invest in Quality Equipment: Use high-quality, modern processing equipment. Well-calibrated oil presses, solvent extractors and refineries are essential to guarantee the quality of processed products.

Use quality raw materials: the quality of oilseeds has a direct impact on the quality of finished products. Make sure seeds are harvested at the right time and are free from impurities.

Comply with food safety standards: Make sure you comply with all food safety standards when processing oilseeds. Food safety is crucial to protecting consumer health and maintaining customer confidence.

Adopt Sustainable Practices: Use sustainable processing practices that minimize environmental impact. For example, recycle processing by-products and use renewable energy sources.

Market research: Before launching a new processed product, carry out market research to understand consumer needs and preferences. This will help you develop products that meet market expectations.

Constantly innovate: Be open to innovation and constantly seek to improve your products and transformation processes. Innovation is essential to staying competitive and responding to market trends.

Collaborate with Experts: Work with oilseed processing experts, nutritionists and marketing specialists to improve the quality of your products and market them effectively.

Focus on Quality: Quality must be a top priority in oilseed processing. High-quality products are essential for building customer loyalty and a good reputation.

Product diversification: Diversify your product range to cater for different market segments. Offer products for different consumption occasions and for different consumer groups.

Use modern technologies: Invest in modern technologies to improve processing efficiency and quality. Modern equipment enables more precise and efficient processes.

Successful Case Studies

To illustrate the possibilities offered by oilseed processing, let's look at a few notable success stories:

Mato Grosso Soybean Oil Extraction Plant, Brazil: This plant uses advanced solvent extraction techniques to produce high-quality soybean oil. By investing in modern equipment and complying with food safety standards, the plant has been able to penetrate international vegetable oil markets, responding to growing consumer demand for high-quality products.

Sofia Sunflower Processing Project, Bulgaria: This community-based project has transformed local sunflower seeds into vegetable oil, margarine and vegetable protein. By providing processing equipment and training to farmers, the project has improved food security and income for rural families. Processed products are sold on local markets and in larger towns, generating additional income for rural families and contributing to the region's economic stability.

La Coopérative de Transformation de Colza de Dijon, France: This cooperative has made a name for itself by producing high-quality rapeseed oil for local and international markets. By using sustainable processing methods and complying with food safety standards, the cooperative has succeeded in increasing production while reducing its environmental impact. It also produces vegetable proteins from rapeseed meal, thus diversifying its sources of income.

These examples show how oilseed processing can be an effective way of improving farm incomes, creating jobs and promoting local economic development. By adopting good practices and taking advantage of market opportunities, farmers can process their oilseed products profitably and sustainably.

Chapter 7: Brewing and winemaking

Brewing and winemaking are two of the oldest food processing techniques, used to produce alcoholic beverages: beer and wine. These processes combine tradition, science and art, and require in-depth knowledge and specific equipment. This chapter explores in detail brewing and winemaking techniques, the equipment required, and the key stages in these processes.

Importance of Brewing and Vinification

Beer and wine production is not only an age-old tradition, it's also a thriving industry. Here's why brewing and winemaking are so important:

Economic value: Beer and wine are among the most widely consumed beverages in the world. Their production generates significant revenues for farmers, breweries, vineyards and producing regions.

Culture and Tradition: Brewing and winemaking are an integral part of many cultures and traditions. They play a central role in festivals, ceremonies and social customs.

Product diversity: Brewing and winemaking create a wide range of products with unique characteristics. Each beer or wine has its own flavor, color and aroma profile.

Innovation and creativity: These processes offer many opportunities for innovation and creativity. Producers can experiment with different recipes, techniques and ingredients to create distinctive products.

Brewing techniques

Brewing is the process of transforming malted barley (or other cereals) into beer. There are several key stages in this process, each of which is essential to the production of high-quality beer.

Malting

Germination: Barley grains are soaked in water and left to germinate. Germination activates the enzymes needed to convert starch into fermentable sugar.

Kilning: After germination, the grains are dried in a kiln. The temperature and duration of kilning influence the color and flavor of the malt.

Brewing

Crushing: The dry malt is crushed to release the starches while leaving the hulls intact, which facilitates filtration.

Mashing: The crushed malt is mixed with hot water in a mashing tun. Enzymes convert the starch into fermentable sugars, forming the wort.

Filtration: The wort is separated from the grain residue (spent grain) by filtration. The spent grains can be used as animal feed.

Boiling: The filtered wort is brought to the boil in a brewing vessel. Hops are added for bitterness, flavor and aroma. Boiling sterilizes the wort and extracts the aromatic compounds from the hops.

Cooling: After boiling, the wort is rapidly cooled to an optimal temperature for fermentation.

Fermentation

Inoculation: The cooled wort is transferred to a fermentation tank and inoculated with yeast. The yeast converts the fermentable sugars into alcohol and carbon dioxide.

Primary fermentation: This phase generally lasts from a few days to a week, during which most of the fermentation takes place.

Secondary fermentation: The beer is transferred to another tank for secondary fermentation. This stage develops the flavors and clarifies the beer.

Packaging

Filtration and carbonation: The beer is filtered to remove yeast residues and other particles. It is then carbonated, either naturally by refermentation, or by adding carbon dioxide.

Bottling: Beer is packaged in bottles, cans or kegs. It can be pasteurized to extend its shelf life.

Wine-making techniques

Winemaking is the process of transforming grapes into wine. Like brewing, this process involves several essential steps to obtain a quality wine.

Harvest

Grape harvest: The grapes are harvested at optimum maturity, either by hand or by machine. The timing of the harvest is crucial to the quality of the wine.

Preparation and pressing

Destemming and crushing: The bunches of grapes are destemmed (separated from the stalks) and crushed to release the juice.

Pressing: The grape juice, called must, is extracted by pressing. The type of pressing (gentle or vigorous) influences the composition of the must and the quality of the wine.

Fermentation

Alcoholic fermentation: The must is transferred to fermentation tanks. Yeast converts sugars into alcohol and carbon dioxide. The temperature and duration of fermentation are controlled to influence the wine's flavor profile.

Malolactic fermentation: After alcoholic fermentation, some wines undergo malolactic fermentation. This stage, carried out by lactic acid bacteria, transforms malic acid into lactic acid, softening the wine.

Maturation and Clarification

Ageing: Young wines are aged in vats, barrels or foudres. Aging allows the wine to develop its aromas and become more refined. Aging conditions (type of container, duration, temperature) strongly influence the character of the wine.

Clarification: Wine is clarified to remove suspended particles. This stage may include filtration, fining (addition of substances to precipitate impurities) and racking (transfer of the clear wine to another container).

Bottling

Stabilization and Final Filtration: Before bottling, the wine can be stabilized (to prevent unwanted fermentation) and filtered one last time to ensure clarity.

Bottling: The wine is bottled, sealed with corks or screw caps. It may be aged in bottle before release to allow the aromas to develop further.

Necessary equipment

Brewing and winemaking require specific equipment adapted to each stage of the process. Here are the main pieces of equipment required for each technique:

Brewing equipment

Malt mill: Used to grind malt into particles of the right size for mashing.

Mash tun: Used to mix ground malt with hot water and activate enzymes.

Filtration tank (Lauter Tun): Used to separate wort from spent grains.

Kettle: Used to boil wort and add hops.

Heat exchanger: Used to cool the wort after boiling.

Fermentation tanks: Tanks where primary and secondary fermentation take place.

Filtration system: Used to clarify beer before packaging.

Packaging equipment: machines for filling bottles, cans or drums.

Wine-making equipment

Crusher-Stemmer: Used to separate grapes from stalks and crush them.

Grape press: Used to extract juice from grapes.

Fermentation tanks: Tanks for alcoholic and malolactic fermentation.

Foudres and Barriques: Containers for maturing wine.

Clarification system: filtration, fining and racking equipment.

Bottling: Machines for filling and sealing wine bottles.

Key steps and best practices

To succeed in brewing and winemaking, it's crucial to follow key steps and respect certain good practices. Here are a few practical tips:

Temperature control: Temperature is a crucial factor in both processes. Carefully control temperature during mashing, fermentation and aging to ensure optimal results.

Ingredient quality: Use high-quality raw materials. For beer, this means fresh malt, aromatic hops and active yeast. For wine, this means ripe, healthy grapes.

Rigorous hygiene: Maintain strict hygiene at every stage to avoid contamination and infection by unwanted micro-organisms.

Monitoring and control: Monitor and control every step of the process. Use measuring instruments to monitor densities, temperatures and pH levels.

Documentation: Keep detailed records of each brew or vintage. Documentation helps you identify successes and areas for improvement.

Training and Expertise: Invest in ongoing training and work with experts to improve your techniques and keep up to date with industry innovations.

Innovation and experimentation: Don't be afraid to innovate and experiment with new recipes, techniques and ingredients to create distinctive products.

Successful Case Studies

To illustrate the possibilities offered by brewing and winemaking, let's look at a few notable success stories:

Munich Craft Brewery, Germany: This craft brewery has combined tradition and innovation to produce high-quality beers. Using traditional brewing techniques and local hops, the brewery has created a range of award-winning beers that are appreciated both locally and internationally.

Domaine Viticole de Napa Valley, USA: This winery has invested in state-of-the-art winemaking techniques to produce world-renowned wines. Using controlled fermentation methods and aging in French oak barrels, the estate has produced wines that have received rave reviews and prestigious awards.

La Microbrasserie de Québec, Canada: This microbrewery has made a name for itself by producing innovative craft beers. By experimenting with local ingredients and unique fermentation techniques, the microbrewery has created distinctive beers that appeal to beer lovers the world over.

Bordeaux Vineyards, France: This historic vineyard has maintained traditional winemaking practices while embracing modern innovations. By combining sustainable viticulture techniques with precise winemaking, the vineyard has produced wines that reflect Bordeaux's unique terroir and are recognized for their exceptional quality.

These examples show how brewing and winemaking can be an effective way of improving farm incomes, creating high-quality products and promoting local economic development. By adopting good practices and taking advantage of market opportunities, producers can transform their raw materials into alcoholic beverages in a profitable and sustainable way.

Chapter 8: Introduction to Chemical Processing

The chemical processing of agricultural products is a fascinating and rapidly expanding field, offering infinite possibilities for adding value to agricultural raw materials. By combining the principles of chemistry with natural resources, it is possible to create a diverse range of products that go far beyond food. This chapter explores the various possibilities for the chemical transformation of agricultural products, highlighting the importance of innovation and sustainability in this sector.

Importance of chemical processing of agricultural products

Chemical processing of agricultural products is crucial for a number of reasons. Not only does it diversify the sources of income for farmers and agri-food industries, it also plays a key role in the transition to a more sustainable, environmentally-friendly economy. Here's why this transformation is so important:

Product diversification: Chemical processing converts agricultural raw materials into a variety of products, including biofuels, cosmetics, pharmaceuticals, bioplastics and many others. This diversification contributes to the economic stability and resilience of agricultural value chains.

Technological innovation: The chemical processing sector is a driver of technological innovation. By developing new processes and applications, it is possible to optimize the use of agricultural resources and reduce waste.

Environmental sustainability: Sustainable chemical transformations can help reduce the ecological footprint of agriculture and industry. For example, the production of biofuels and bioplastics from renewable agricultural materials helps reduce greenhouse gas emissions and dependence on fossil fuels.

Upgrading agricultural waste: Chemical processing offers opportunities for upgrading agricultural residues and waste, transforming by-products often regarded as nuisances into valuable resources.

Chemical transformation possibilities

Agricultural produce can be chemically transformed into a multitude of useful products. Here are some of the main chemical transformation routes, each with its own specific applications and benefits.

Biofuels

Biofuels are fuels produced from renewable organic matter. They represent a sustainable alternative to fossil fuels and play a key role in reducing greenhouse gas emissions.

Bioethanol: Bioethanol is produced by fermenting sugars from crops such as corn, sugar cane and sugar beet. It is commonly used as a gasoline additive to improve combustion and reduce pollutant emissions.

Biodiesel: Biodiesel is produced by the transesterification of vegetable oils and animal fats. Common feedstocks include soybean, rapeseed and used cooking oil. Biodiesel can be used in conventional diesel engines and is biodegradable and non-toxic.

Biogas: Biogas is produced by the anaerobic digestion of biomass, including agricultural waste, crop residues and livestock effluents. It is composed mainly of methane and carbon dioxide, and can be used as a renewable energy source for the production of heat, electricity and vehicle fuel.

Cosmetics

Agricultural products can be transformed into a variety of natural cosmetics, meeting growing consumer demand for safe, environmentally-friendly products.

Essential oils: Essential oils are extracted from aromatic plants such as lavender, peppermint and eucalyptus. They are used in personal care products for their aromatic and therapeutic properties.

Vegetable oils: Vegetable oils, such as argan oil, jojoba oil and coconut oil, are commonly used in skin and hair care for their moisturizing and nourishing properties.

Botanical extracts: Plant extracts such as aloe vera, green tea and chamomile are used for their soothing, anti-inflammatory and antioxidant properties. They are incorporated into a range of cosmetic products, including creams, lotions and serums.

Pharmaceuticals

Agricultural produce is a rich source of bioactive compounds that can be transformed into active pharmaceutical ingredients to treat a variety of diseases and conditions.

Alkaloids: Alkaloids are nitrogen-containing organic compounds produced by certain plants. Morphine, quinine and nicotine are examples of alkaloids used in medicine.

Flavonoids: Flavonoids are powerful antioxidants found in many fruits and vegetables. They have anti-inflammatory, antiviral and anticancer properties, and are used in food supplements and medicines.

Saponins: Saponins are triterpene or steroidal glycosides found in plants such as soy and quinoa. They are used for their cholesterol-lowering and immunostimulant properties.

Bioplastics

Bioplastics are plastic materials produced from renewable sources, such as corn starch, sugarcane and cellulose. They represent an environmentally-friendly alternative to traditional petroleum-derived plastics.

PLA (Polylactic Acid): PLA is a bioplastic produced by fermenting sugars from crops such as corn. It is used to manufacture packaging, bags, cups and disposable utensils.

PHA (Polyhydroxyalkanoates): PHAs are bioplastics produced by bacteria fermenting sugars or vegetable oils. They are used in medical applications, food packaging and biodegradable products.

Thermoplastic starch: Thermoplastic starch is produced by modifying starch extracted from crops such as corn and potatoes. It is used to make biodegradable films, bags and cushioning materials.

Cleaning Products and Detergents

Agricultural products can also be transformed into ingredients for eco-friendly cleaning products and detergents.

Plant-derived surfactants: Plant-derived surfactants, such as glucosides and lecithins, are used for their foaming and cleansing properties in soaps and detergents.

Vegetable solvents: Solvents derived from agriculture, such as ethanol and limonene, are used in cleaning products to dissolve grease and dirt.

Natural bleaching agents: Natural bleaching agents, such as sodium percarbonate, are derived from agricultural materials and used for their cleaning and disinfecting properties.

Importance of Innovation and Sustainability

Innovation and sustainability are two essential pillars for success in the chemical processing of agricultural products. Here's why they are so crucial:

Meeting market requirements: Consumers and regulators are increasingly demanding sustainable, safe and environmentally-friendly products. Innovation makes it possible to develop products that meet these expectations while offering competitive advantages.

Resource optimization: Technological innovation optimizes the use of agricultural resources, reducing waste and improving the efficiency of transformation processes. This contributes to economic and environmental sustainability.

Reducing Environmental Impact: Sustainable chemical transformation processes minimize environmental impact by reducing greenhouse gas emissions, using renewable raw materials and promoting product recyclability.

Health and safety: Innovation in chemical processing enables us to develop safer products for consumers and workers. This includes reducing the use of toxic substances and developing cleaner processes.

Local economic development: By promoting local processing of agricultural products, innovation and sustainability contribute to the economic development of rural areas. This creates jobs, increases farmers' incomes and supports local economies.

Practical Tips for Successful Chemical Processing

To succeed in the chemical processing of agricultural products, it's important to follow certain good practices and focus on key aspects. Here are a few tips to maximize your chances of success:

Investing in Research and Development: Research and development (R&D) are essential to innovate and improve chemical transformation processes. Invest in R&D projects to discover new applications and optimize existing processes.

Collaborate with Experts: Work with experts in chemistry, agriculture and engineering to improve product quality and develop sustainable processes. Partnerships with research institutes and universities can also be beneficial.

Use Quality Raw Materials: The quality of raw materials directly influences the quality of finished products. Make sure your agricultural raw materials are of the highest quality and sustainably grown.

Comply with safety and environmental standards: Make sure you comply with all safety and environmental standards during chemical processing. Regulatory compliance is crucial to protect the health of workers, consumers and the environment.

Adopt Sustainable Practices: Use sustainable processing practices that minimize environmental impact. This includes using green solvents, reducing waste and optimizing energy efficiency.

Market research: Before launching a new processed chemical product, carry out market research to understand consumer needs and preferences. This will help you develop products that meet market expectations.

Constantly innovate: Be open to innovation and constantly seek to improve your products and transformation processes. Innovation is essential to staying competitive and responding to market trends.

Focus on Quality: Quality must be a top priority in the chemical processing of agricultural products. High-quality products are essential for building customer loyalty and a good reputation.

Product diversification: Diversify your product range to meet the needs of different market segments. Offer products for different applications and consumer groups.

Use modern technologies: Invest in modern technologies to improve processing efficiency and quality. Modern equipment enables more precise and efficient processes.

Successful Case Studies

To illustrate the possibilities offered by chemical processing of agricultural products, let's take a look at a few notable success stories:

Bologna Bioplastics Plant, Italy: This plant uses corn starch to produce PLA bioplastics. By investing in state-of-the-art fermentation and polymerization technologies, the plant has succeeded in creating biodegradable bioplastics used in food packaging, bags and disposable utensils. These bioplastics meet the growing demand for sustainable alternatives to traditional plastics.

São Paulo Bioethanol Production Project, Brazil: This project uses sugarcane to produce bioethanol, a renewable biofuel. Using advanced fermentation and distillation technologies, the project has succeeded in optimizing bioethanol production while reducing greenhouse gas emissions. The bioethanol produced is used as a gasoline additive, helping to reduce dependence on fossil fuels.

La Coopérative de Transformation d'Huiles Essentielles de Provence, France: This cooperative transforms local aromatic plants such as lavender and rosemary into high-quality essential oils. Using environmentally-friendly steam distillation techniques, the cooperative has succeeded in producing essential oils used in cosmetics, perfumes and therapeutic products. The cooperative's products are exported worldwide, generating significant income for its members.

Bavaria Biofuel Plant, Germany: This plant converts crop residues and agricultural waste into biogas, which is used to generate electricity and heat. Using advanced anaerobic digestion technologies, the plant has successfully valorized agricultural waste while contributing to the production of renewable energy. The project has reduced methane emissions and created local jobs in the region.

These examples show how chemical processing of agricultural products can be an effective way of improving farm incomes, creating innovative products and promoting sustainable development. By adopting good practices and taking advantage of market opportunities, farmers and processors can transform their raw materials into chemical products in a profitable and sustainable way.

Chapter 9: Conversion to biofuels

Biofuels represent a sustainable alternative to traditional fossil fuels. They are produced from renewable agricultural materials and offer significant economic and environmental benefits. This chapter explores in detail the process of producing biofuels from agricultural materials, their environmental impact, and the economic prospects of this transformation.

Importance of biofuels

Biofuels are essential for a number of reasons. They help reduce greenhouse gas emissions, lessen dependence on fossil fuels and provide a new source of income for farmers. Here's why biofuels are crucial:

Reduced greenhouse gas emissions: Biofuels produce less CO_2 and other pollutants than fossil fuels, helping to combat climate change.

Reduced dependence on fossil fuels: Biofuels help to diversify energy sources and reduce dependence on oil imports, thus improving energy security.

New income for farmers: Biofuel production offers a new source of income for farmers by adding value to agricultural raw materials and crop residues.

Local economic development: Biofuel production can stimulate local economic development by creating jobs in rural areas and promoting investment in infrastructure.

Biofuel Production Process

Biofuel production involves several stages, each requiring specific techniques and adapted equipment. Here are the main types of biofuel and the associated production processes:

Bioethanol

Bioethanol is a liquid biofuel produced by the fermentation of sugars present in agricultural materials such as corn, sugar cane and sugar beet. Here are the key stages in its production:

Raw material preparation: Raw materials are cleaned and prepared for fermentation. In the case of corn, the kernels are ground to release the starch.

Saccharification: Starch is converted into fermentable sugars by the addition of specific enzymes. This process is called saccharification.

Fermentation: Sugars are fermented by yeast to produce ethanol and carbon dioxide. Fermentation generally lasts 48 to 72 hours.

Distillation: The fermented mixture, called wort, is distilled to separate the ethanol from the other components. Pure ethanol is collected in vapor form, then condensed to a liquid.

Dehydration: Distilled ethanol still contains a small amount of water. It is dehydrated to obtain anhydrous bioethanol, which is then blended with gasoline to form fuels such as E85 (85% ethanol, 15% gasoline).

Biodiesel

Biodiesel is a liquid biofuel produced by the transesterification of vegetable oils and animal fats. Here are the key stages in its production:

Raw materials preparation: Vegetable oils (e.g. soybean, rapeseed or sunflower oil) or animal fats are filtered to remove impurities.

Transesterification: Oils are mixed with an alcohol (usually methanol) and a catalyst (often sodium hydroxide) to produce methyl esters (biodiesel) and glycerol as a by-product.

Separation: The reaction mixture is left to stand to allow separation of the biodiesel from the glycerol. The biodiesel floats to the surface and is recovered.

Purification: Raw biodiesel is washed with water to remove traces of methanol, catalyst and soap. It is then dried to remove any residual water.

Packaging: Purified biodiesel is packaged and can be used directly in diesel engines or blended with fossil diesel.

Biogas

Biogas is a gaseous fuel produced by the anaerobic digestion of biomass, including agricultural waste, crop residues and livestock effluents. Here are the key stages in its production:

Raw materials collection: Biomass is collected and transported to an anaerobic digestion plant.

Pre-treatment: Biomass is pre-treated to optimize biological degradation. This may include grinding, mixing and adjusting moisture and pH.

Anaerobic digestion: Biomass is placed in an anaerobic digester, where it is broken down by micro-organisms in the absence of oxygen. This process produces biogas (mainly methane and carbon dioxide) and digestate (a nutrient-rich residue).

Biogas capture and storage: Biogas is captured and stored in tanks. It can be used to generate heat and electricity, or purified for use as a vehicle fuel.

Digestate use: Digestate is a nutrient-rich by-product that can be used as an organic fertilizer to improve soil fertility.

Environmental impact of biofuels

Biofuels offer a number of environmental advantages over fossil fuels. However, their production and use can also have negative impacts if they are not managed sustainably. Here are some of the environmental impacts of biofuels:

Reduced greenhouse gas emissions: Biofuels produce less CO2 and other pollutants than fossil fuels. For example, the use of biodiesel can reduce CO2 emissions by 50 to 80% compared with fossil diesel.

Use of agricultural waste: Biogas production from agricultural waste and crop residues contributes to sustainable waste management and the reduction of methane emissions from landfills and livestock effluents.

Impact on land use: The production of energy crops for biofuels can lead to changes in land use, including the conversion of agricultural land, forests and natural areas. This can affect biodiversity and the availability of land for food production.

Water consumption: The production of biofuels, particularly bioethanol, can require large quantities of water for crop irrigation and processing. Sustainable water management is essential to minimize the impact on water resources.

Impact on air and water quality: The production and use of biofuels can have impacts on air and water quality. For example, the intensive cultivation of raw materials for biofuels can lead to pollution from pesticides and fertilizers.

Economic outlook for biofuels

Biofuels offer promising economic prospects for farmers, processors and local economies. Here are some of the main economic prospects for biofuels:

Job creation: Biofuel production creates jobs at various stages of the value chain, from the cultivation of raw materials to the processing, packaging and distribution of finished products.

Diversification of farm income: Biofuels offer a new source of income for farmers by adding value to agricultural raw materials and crop residues. This contributes to the economic stability and resilience of farms.

Investment Attraction: The biofuels sector attracts investment in infrastructure, technology and R&D projects. This stimulates local and regional economic development.

Reduced energy costs: The use of biofuels can reduce energy costs for local farms and industries. For example, biogas can be used to generate electricity and heat on site, reducing energy costs.

Access to international markets: Biofuels offer export opportunities to international markets, where demand for renewable and sustainable energy sources is growing. This opens up new commercial outlets for biofuel producers.

Practical Tips for Successful Biofuel Production

To succeed in biofuel production, it's important to follow certain best practices and focus on key aspects. Here are a few tips to maximize your chances of success:

Choosing the right feedstocks: Select feedstocks suited to your region and infrastructure. For example, corn and sugarcane are common feedstocks for bioethanol, while soy and rapeseed are commonly used for biodiesel.

Optimize Production Processes: Invest in modern, efficient technologies to optimize biofuel production processes. This includes the use of specific enzymes for saccharification, efficient catalysts for transesterification and high-performance anaerobic digesters for biogas production.

Comply with environmental standards: Make sure you meet all environmental standards when producing biofuels. Regulatory compliance is crucial to protecting the environment and maintaining consumer confidence.

Adopt Sustainable Practices: Use sustainable production practices that minimize environmental impact. For example, recycle production by-products, use renewable raw materials and manage water and soil resources responsibly.

Market research: Before launching biofuel production, carry out market research to understand market demand and trends. This will help you develop products that meet consumer expectations and identify business opportunities.

Constantly innovate: Be open to innovation and constantly seek to improve your production processes and develop new products. Innovation is essential to staying competitive and responding to market trends.

Collaborate with Experts: Work with experts in biotechnology, engineering and agriculture to improve product quality and optimize production processes. Partnerships with research institutes and universities can also be beneficial.

Focus on Quality: Quality must be a top priority in biofuel production. High-quality products are essential for building customer loyalty and a good reputation.

Product diversification: Diversify your product range to meet the needs of different market segments. Offer biofuels for different applications, such as bioethanol for vehicles, biodiesel for diesel engines and biogas for power generation.

Use Modern Technologies: Invest in modern technologies to improve the efficiency and quality of biofuel production. Modern equipment enables more precise and efficient processes.

Successful Case Studies

To illustrate the possibilities offered by biofuel production, let's look at a few notable success stories:

Mato Grosso Bioethanol Plant, Brazil: This plant uses sugarcane to produce high-quality bioethanol. By investing in state-of-the-art fermentation and distillation technologies, the plant has succeeded in optimizing bioethanol production while reducing greenhouse gas emissions. The bioethanol produced is used as a gasoline additive, helping to reduce dependence on fossil fuels.

The Bavarian Biogas Project, Germany: This community project converts crop residues and agricultural waste into biogas, which is used to generate electricity and heat. Using advanced anaerobic digestion technologies, the project has successfully valorized agricultural waste while contributing to the production of renewable energy. The project has reduced methane emissions and created local jobs in the region.

Saskatchewan Biodiesel Cooperative, Canada: This cooperative transforms local canola oils into high-quality biodiesel. Using efficient and environmentally-friendly transesterification methods, the cooperative has succeeded in producing biodiesel used in local diesel engines and exported to international markets. The project has improved the incomes of member farmers and contributed to the energy sustainability of the region.

These examples show how biofuel production can be an effective way of improving farm incomes, creating high-quality products and promoting sustainable development. By adopting good practices and taking advantage of market opportunities, farmers and processors can transform their raw materials into biofuels in a profitable and sustainable way.

Chapter 10: Transformation into Cosmetic Products

Processing agricultural materials into cosmetics is a promising way of adding value to crops and diversifying farmers' sources of income. Cosmetic products, such as creams, lotions and essential oils, are in great demand and offer significant economic opportunities. This chapter explores techniques for transforming agricultural materials into cosmetics, regulatory aspects and market prospects.

Importance of Cosmetics Processing

Processing into cosmetic products offers several advantages for farmers and the food industry:

Added value: Processing agricultural materials into cosmetics adds value to raw materials, offering higher profit margins.

Income diversification: By producing cosmetics, farmers can diversify their sources of income, reducing their dependence on a single crop or market.

Meeting consumer demand: Demand for natural and organic cosmetics is growing, creating opportunities for products derived from agricultural materials.

Promoting sustainability: Natural and organic cosmetics are often perceived as being more respectful of the environment, which can enhance the reputation of producers and attract sustainability-conscious consumers.

Techniques for Processing Agricultural Raw Materials into Cosmetic Products

The transformation of agricultural materials into cosmetics involves several stages, from the harvesting of raw materials to the formulation of finished products. Here's an overview of the main techniques used:

Essential oil extraction

Essential oils are concentrated extracts of aromatic plants, used for their therapeutic properties and fragrances. Here are the main extraction methods:

Steam distillation: The most common method for extracting essential oils. Plants are exposed to steam, releasing the volatile oils, which are then condensed and separated from the water.

Cold Pressing: Used mainly for citrus fruits. Fruit peels are mechanically pressed to extract the essential oils.

Solvent extraction: Used for delicate plants. Plant material is soaked in a solvent (such as ethanol) which dissolves the essential oils. The solvent is then evaporated to leave the pure oils.

Supercritical CO2 extraction: An advanced method using high-pressure carbon dioxide to extract essential oils without heat. This better preserves delicate aromatic compounds.

Production of Vegetable Oils

Vegetable oils are used in many cosmetic products for their moisturizing and nourishing properties. Here are the main stages in their production:

Harvesting and drying: Oil seeds or fruits are harvested and dried to reduce their water content.

Cold Pressing: Seeds or fruits are mechanically cold-pressed to extract the oil. This method preserves the oils' nutrients and beneficial properties.

Filtration: The crude oil is filtered to remove impurities and obtain a clear, pure oil.

Production of creams and lotions

Creams and lotions are emulsions of oil and water, often enriched with plant extracts and essential oils. Here's how they're made:

Formulation: Development of the formula by mixing vegetable oils, water, emulsifiers and other active ingredients.

Mixing and heating: Ingredients are mixed and heated to form a stable emulsion. Temperature and mixing time are controlled to ensure a uniform texture.

Cooling: The emulsion is cooled slowly to stabilize the texture and prevent phase separation.

Packaging: Creams and lotions are packaged for sale in jars, tubes or bottles.

Incorporation of Botanical Extracts

Botanical extracts are added to cosmetic products for their beneficial properties for skin and hair. Here's how they are prepared and used:

Extraction: Plants are soaked in a solvent (water, alcohol, glycerine) to extract the active compounds.

Concentration: Extracts are concentrated by evaporation of the solvent, if necessary, to increase the content of active compounds.

Incorporation: Concentrated extracts are added to formulas for creams, lotions, shampoos, etc., respecting recommended dosages for efficacy and safety.

Cosmetics regulations

The production and marketing of cosmetic products are strictly regulated to ensure consumer safety and product quality. Here are the main regulatory aspects to consider:

Permitted ingredients: Regulations specify the ingredients authorized and prohibited in cosmetic products. It is important to check the conformity of ingredients used in formulas.

Labeling: Cosmetic products must be properly labeled, including a complete list of ingredients, instructions for use, warnings and manufacturer information.

Safety tests: Cosmetic products must pass safety tests to ensure that they do not cause irritation, allergies or other undesirable effects. These tests include stability tests, skin compatibility tests and microbiological tests.

Product registration: In some countries, cosmetic products must be registered with regulatory authorities before they can be marketed. This includes the submission of dossiers detailing composition, manufacturing methods and safety test results.

Good Manufacturing Practices (GMP): Cosmetics manufacturers must follow GMP to ensure product quality and safety. This includes quality control of raw materials, control of production processes and documentation of procedures.

Cosmetics market

The cosmetics market is vast and constantly evolving. It offers significant opportunities for agricultural producers wishing to engage in the transformation of raw materials into beauty products. Here are some market trends and prospects:

Growing demand for natural and organic products: Consumers are increasingly aware of the importance of natural and organic ingredients in cosmetics. This creates a demand for products derived from pesticide- and chemical-free agricultural materials.

Innovation and diversification: Innovations in formulations and transformation processes offer opportunities to develop unique and differentiated cosmetic products. Producers can diversify their product range to meet the varied needs of consumers.

Market segmentation: The cosmetics market is segmented into different categories, such as skin care, hair care, make-up products and so on. Each segment offers specific opportunities based on consumer trends and preferences.

Distribution and Sales Channels: Cosmetic products can be sold through a variety of channels, including retail stores, pharmacies, online stores, beauty salons, etc. Diversifying sales channels enables us to reach a wider audience and optimize sales.

Export: Natural and organic cosmetics are in high demand on international markets. Producers can explore export opportunities to increase their revenues and extend their reach.

Practical Tips for Successful Cosmetics Processing

To succeed in transforming agricultural materials into cosmetics, it's important to follow certain best practices and focus on key aspects. Here are a few tips to maximize your chances of success:

Invest in Quality: The quality of raw materials and finished products is essential to success in the cosmetics market. Use high-quality ingredients, follow good manufacturing practices and carry out rigorous safety tests.

Innovate and Differentiate: Innovation is crucial to standing out in a competitive market. Develop unique and differentiated products using innovative formulations and integrating natural and organic ingredients.

Comply with regulations: Make sure you comply with all local and international cosmetics regulations. Regulatory compliance is essential to guarantee consumer safety and avoid legal problems.

Market research: Before launching a new product, carry out market research to understand consumer trends, needs and preferences. This will help you develop products that meet market expectations.

Collaborate with Experts: Work with experts in cosmetic formulation, regulation and marketing to improve product quality and optimize your production processes. Partnerships with research laboratories and beauty institutes can also be beneficial.

Marketing and communication: Marketing and communication are essential to success in the cosmetics market. Develop an effective marketing strategy to promote your products and attract consumers. Use social media, advertising campaigns and beauty events to increase your brand's visibility.

Use modern technologies: Invest in modern technologies to improve production efficiency and quality. Modern equipment enables more precise and efficient processes.

Product diversification: Diversify your product range to cater for different market segments. Offer products for different applications, such as skincare, hair care, make-up, etc.

Support Sustainability: Adopt sustainable production practices that minimize environmental impact. Use renewable ingredients, recycle production waste and reduce the carbon footprint of your operations.

Creating a Trusted Brand: Consumer trust is essential in the cosmetics market. Develop a trusted brand by guaranteeing the quality, safety and transparency of your products. Communicate openly about the ingredients, manufacturing processes and benefits of your products.

Successful Case Studies

To illustrate the possibilities offered by the transformation of agricultural materials into cosmetic products, let's look at a few notable success stories:

L'Entreprise de Cosmétiques Naturels de Provence, France: This company uses local aromatic plants, such as lavender, rosemary and thyme, to produce essential oils and natural creams. By focusing on quality ingredients and sustainable production practices, the company has made a name for itself in the natural cosmetics market. Products are sold in specialized boutiques and online, attracting an environmentally conscious clientele.

La Coopérative de Beauté d'Argan, Morocco: This cooperative brings together rural women who harvest argan nuts to produce high-quality argan oil. Using traditional cold-pressing methods, the cooperative produces a nutrient-rich oil used in skin and hair care. The cooperative's products are exported to international markets, generating significant income for members and supporting local development.

The Botanical Skincare Brand from California, USA: This skincare brand uses botanical extracts from organic farming to formulate innovative beauty products. By integrating modern formulation technologies and focusing on the purity of ingredients, the brand has succeeded in attracting a loyal customer base. Products are distributed in high-end retail stores and via online boutiques, with communication focused on product transparency and efficacy.

These examples show how transforming agricultural materials into cosmetics can be an effective way of improving farm incomes, creating high-quality products and promoting sustainable development. By adopting good practices and taking advantage of market opportunities, farmers and processors can transform their raw materials into cosmetics in a profitable and sustainable way.

Chapter 11: Transformation into Pharmaceutical Products

The transformation of agricultural products into Active Pharmaceutical Ingredients (APIs) is a cutting-edge field that combines agriculture and medical science to produce medicines and supplements essential to human health. This transformation requires scientific rigor and strict compliance with quality standards to guarantee the efficacy and safety of pharmaceutical products. This chapter explores the different methods of transforming agricultural products into APIs, the importance of quality and production standards, and the economic prospects of this transformation.

Importance of Pharmaceutical Processing

The transformation of agricultural products into active pharmaceutical ingredients is crucial for several reasons:

Diversification of sources of income: This enables farmers to diversify their sources of income by supplying raw materials for the pharmaceutical industry.

Medical innovation and progress: Agricultural products offer a multitude of bioactive compounds that can be used to develop new drugs and treatments.

Promoting public health: Active pharmaceutical ingredients derived from plants and other agricultural products can contribute to public health by providing natural, effective alternatives to synthetic medicines.

Sustainable development: Using renewable resources for drug production contributes to environmental sustainability and reduces dependence on fossil fuels.

Pharmaceutical Transformation Processes

The transformation of agricultural products into active pharmaceutical ingredients involves several key stages, from the cultivation of raw materials to the formulation of finished products. Here's an overview of the main stages in the process:

Growing and harvesting raw materials

Medicinal plant selection: Medicinal plants are selected on the basis of their content of bioactive compounds. The selection includes plants such as foxglove (source of digitalis), opium poppy (source of morphine) and ginseng (used for its adaptogenic properties).

Optimal Growing Conditions: Growing conditions, such as soil, climate and agricultural practices, are optimized to maximize the concentration of bioactive compounds. This may include the use of organic methods to avoid pesticide residues.

Harvesting at the right time: Medicinal plants are harvested at the optimum time to ensure maximum content of active ingredients. For example, some plants are harvested while in flower, while others are harvested when the roots or leaves are richest in active compounds.

Extraction and Purification of Bioactive Compounds

Extraction: Bioactive compounds are extracted from plants using solvents (water, alcohol, ether) or more advanced methods such as supercritical CO2 extraction. The aim is to obtain a concentrated extract containing the active ingredients.

Purification: crude extracts are purified to isolate specific active ingredients. This may include techniques such as column chromatography, distillation and recrystallization. Purification is essential to eliminate impurities and undesirable substances.

Identification and Quantification: Purified active ingredients are identified and quantified using analytical techniques such as high-performance liquid chromatography (HPLC) and mass spectrometry. These techniques guarantee the purity and concentration of active ingredients.

Pharmaceutical Formulation

Formulation Development: Active ingredients are incorporated into formulations suitable for their administration. This may include the formulation of tablets, capsules, injectable solutions, creams or syrups.

Stability testing: Formulations are subjected to stability tests to ensure that they maintain their efficacy and safety over the shelf life. This includes tests under different conditions of temperature, humidity and light.

Quality control: Each production batch undergoes rigorous quality controls to verify compliance with purity, concentration and safety specifications. Tests include checking for the absence of microbial contaminants and toxic substances.

Importance of Quality and Production Standards

Quality and production standards are crucial aspects in the transformation of agricultural products into active pharmaceutical ingredients. Here's why they're so important:

Patient safety : Patient safety is our top priority. Medicines must be free from contaminants and impurities that could cause adverse effects.

Therapeutic efficacy: Active ingredients must be present in appropriate concentrations to guarantee therapeutic efficacy. Incorrect concentrations may render the drug ineffective or dangerous.

Regulatory Compliance: Medicines must comply with the strict regulations imposed by health authorities, such as the FDA (Food and Drug Administration) in the USA and the EMA (European Medicines Agency) in Europe. Regulatory compliance is essential to obtain marketing authorization.

Reputation and trust: The quality of pharmaceutical products is essential to maintaining the reputation of manufacturers and the trust of consumers and healthcare professionals.

Good Manufacturing Practices (GMP)

Good Manufacturing Practice (GMP) is a set of guidelines covering all aspects of pharmaceutical production, from the cultivation of raw materials to the manufacture and packaging of finished products. Here are some key aspects of GMP:

Rigorous documentation: All stages of production must be accurately documented to guarantee traceability and compliance with quality standards.

Staff training: Staff must be trained in GMP and company-specific procedures to ensure consistent, high-quality operations.

Plant control : Production facilities must be designed and maintained to minimize the risk of contamination and ensure a clean, controlled production environment.

Raw materials control: Raw materials must be inspected and tested for compliance with specifications before being used in production.

Verification and Validation: Production processes must be verified and validated to ensure that they consistently produce products that meet quality specifications.

Pharmaceutical regulation

Pharmaceutical products are subject to strict regulations to guarantee their safety, efficacy and quality. Here are some key aspects of pharmaceutical regulation:

Marketing Authorization: Medicines must obtain a Marketing Authorization (MA) issued by the relevant health authorities. This authorization is based on preclinical and clinical data demonstrating the drug's safety and efficacy.

Pharmacovigilance: Manufacturers must monitor the safety of medicines after they have been placed on the market, and report any adverse effects to the health authorities. Pharmacovigilance is essential for identifying and managing the risks associated with medicines.

Good Distribution Practices (GDP): GMPs are guidelines covering the distribution of medicines to ensure that they are stored and transported in conditions appropriate to maintain their quality.

Audit and inspection: Health authorities may carry out audits and inspections of production facilities to verify compliance with GMP and other regulatory requirements.

Economic outlook for pharmaceutical processing

The transformation of agricultural products into active pharmaceutical ingredients offers promising economic prospects for farmers, processors and pharmaceutical industries. Here are some of the main economic prospects:

Creating added value: Processing agricultural raw materials into APIs adds significant value to raw products, increasing profit margins and incomes for farmers and processors.

Access to international markets: APIs derived from agricultural products can be exported to international markets, offering opportunities for growth and revenue diversification.

Development of new products: Innovation in the transformation of agricultural products into APIs enables the development of new drugs and treatments, meeting unmet medical needs and opening up new market opportunities.

Support for rural development: Medicinal plant production and processing into APIs can stimulate economic development in rural areas by creating jobs and improving local infrastructure.

Partnerships and Collaborations: Farmers and processors can collaborate with pharmaceutical companies, research institutes and universities to develop advanced processing and innovative products.

Practical Tips for Successful Pharmaceutical Processing

To succeed in transforming agricultural products into active pharmaceutical ingredients, it's important to follow certain best practices and focus on key aspects. Here are a few tips to maximize your chances of success:

Investing in Research and Development (R&D): R&D is essential for discovering new bioactive compounds and developing efficient transformation processes. Invest in R&D projects to stay at the forefront of innovation.

Collaborate with Experts: Work with experts in pharmacology, analytical chemistry and biotechnology to improve the quality of your products and optimize your production processes. Partnerships with research institutes and universities can also be beneficial.

Use quality raw materials: The quality of raw materials has a direct influence on the quality of APIs. Make sure that medicinal plants are grown in optimal conditions and are free from contaminants.

Comply with safety and quality standards: Make sure you comply with all safety and quality standards when producing APIs. Regulatory compliance is crucial to ensuring patient safety and obtaining marketing authorization.

Adopt Sustainable Practices: Use sustainable growing and processing practices that minimize environmental impact. This includes the use of organic methods, responsible resource management and waste reduction.

Market research: Before launching a new pharmaceutical product, carry out market research to understand market needs and trends. This will help you develop products that meet the expectations of consumers and healthcare professionals.

Constantly innovate: Be open to innovation and constantly seek to improve your products and production processes. Innovation is essential to staying competitive and responding to market trends.

Focus on Quality: Quality must be a top priority when producing IPAs. High-quality products are essential for building customer loyalty and a good reputation.

Product diversification: Diversify your product range to cater for different market segments. Offer APIs for different therapeutic applications, such as drugs, dietary supplements and natural health products.

Use modern technologies: Invest in modern technologies to improve production efficiency and quality. Modern equipment enables more precise and efficient processes.

Successful Case Studies

To illustrate the possibilities offered by the transformation of agricultural products into active pharmaceutical ingredients, let's look at a few notable success stories:

L'Entreprise de Phytopharmacie de Madagascar: This company uses local medicinal plants to produce active pharmaceutical ingredients used in treatments for malaria, infections and inflammatory diseases. By collaborating with local and international research institutes, the company has succeeded in developing advanced extraction and purification processes, guaranteeing the quality and efficacy of APIs.

The South Korean Ginseng Cooperative: This cooperative brings together farmers specializing in the cultivation of ginseng, a medicinal plant used for its adaptogenic and immunostimulant properties. Using organic cultivation methods and investing in modern processing technologies, the cooperative produces high-quality ginseng extracts, used in dietary supplements and medicines.

Morphine Processing Plant in India: This plant transforms opium poppies into morphine, an active pharmaceutical ingredient used in the treatment of pain. Meeting strict safety and quality standards, the plant produces pure morphine, used in hospitals and clinics worldwide.

These examples show how transforming agricultural products into active pharmaceutical ingredients can be an effective way of improving farm incomes, creating high-quality products and promoting sustainable development. By adopting good practices and taking advantage of market opportunities, farmers and processors can transform their raw materials into APIs in a profitable and sustainable way.

Chapter 12: Conversion to Biodiesel and Bioethanol

The conversion of specific crops into biofuels such as biodiesel and bioethanol represents a major step forward in the quest for renewable and sustainable energy sources. These biofuels offer an environmentally-friendly alternative to fossil fuels, reducing carbon footprints and improving energy security. This chapter explores in detail the biodiesel and bioethanol production processes, their applications and the associated ecological benefits.

Importance of biofuels

Biofuels play a crucial role in the transition to more sustainable energy sources. Here's why they're important:

Reduced greenhouse gas emissions: Biofuels emit less CO2 and pollutants than fossil fuels, helping to combat climate change.

Energy security: By diversifying energy sources, biofuels reduce dependence on oil imports and enhance energy security.

Rural development: Biofuel production can stimulate rural economic development by creating jobs and generating additional income for farmers.

Waste utilization: Biofuels make it possible to valorize crop residues and agricultural waste, thus reducing pollution and waste.

Biodiesel Production Processes

Biodiesel is a renewable fuel produced from vegetable oils or animal fats. Here are the main stages in its production:

Raw materials

Feedstocks commonly used for biodiesel production include :

Vegetable oils: soybean, rapeseed, sunflower, palm.

Animal fats: Beef, pork and poultry fats.

Used oils: Used cooking oils from restaurants and the food industry.

Oil pre-treatment

Filtration: Oils and greases are filtered to remove solid impurities.

Neutralization: Free fatty acids present in oils can be neutralized by adding base, thus reducing the risk of soap formation during transesterification.

Transesterification

Transesterification is the central chemical process in biodiesel production. It consists in converting triglycerides (components of oils and fats) into methyl esters (biodiesel) and glycerol. Here are the main stages:

Reagent Mixing: Filtered oils are mixed with methanol and a catalyst (often sodium or potassium hydroxide).

Reaction: The mixture is heated to a controlled temperature (approx. 60°C) and stirred to allow the transesterification reaction to take place. This reaction generally takes 1 to 2 hours.

Separation: After the reaction, the mixture is left to stand to allow the phases to separate. The less dense biodiesel floats on top of the glycerol.

Purification: Raw biodiesel is washed with water to remove catalyst, methanol and soap residues. It is then dried to remove all residual water.

Biodiesel applications

Biodiesel can be used in a number of applications, including :

Transport: Used as fuel for diesel-powered vehicles, either pure (B100) or blended with fossil diesel (B20, B5).

Heating: Used as fuel for domestic and industrial heating boilers.

Electricity Generation: Used in diesel generators to produce electricity, particularly in rural and remote areas.

Biodiesel's environmental benefits

Emissions reduction: Biodiesel reduces emissions of CO_2, fine particles, sulfur and polycyclic aromatic hydrocarbons (PAHs).

Biodegradability: Biodiesel is biodegradable and non-toxic, reducing the risk of pollution in the event of a spill.

Waste utilization: Producing biodiesel from used oil enables us to recover waste and reduce its environmental impact.

Bioethanol Production Processes

Bioethanol is a liquid biofuel produced by fermenting sugars present in plant matter. Here are the main stages in its production:

Raw materials

Feedstocks commonly used for bioethanol production include :

Sugar-rich crops: sugar cane, sugar beet.

Starch-rich crops: corn, wheat, sorghum.

Lignocellulosic biomass: wood waste, straw, corn stalks.

Raw materials pre-treatment

Milling and Grinding: The grains are ground to release the starch, while the sugar plants are pressed to extract the sugar-rich juice.

Hydrolysis: In the case of lignocellulosic materials, chemical or enzymatic pretreatment is required to break down cellulose and hemicellulose into fermentable sugars.

Saccharification and Fermentation

Saccharification: Starch is hydrolyzed into fermentable sugars (glucose) using specific enzymes (amylases).

Fermentation: Fermentable sugars are converted into ethanol and carbon dioxide by yeast. Fermentation generally lasts 24 to 48 hours.

Distillation and Dehydration

Distillation: The fermented mixture (beer) is distilled to separate the ethanol from the other components. The crude ethanol obtained is generally 95% pure.

Dehydration: Crude ethanol is dehydrated to remove the remaining water, producing anhydrous ethanol (99.5% purity).

Bioethanol applications

Bioethanol can be used in a number of applications, including :

Vehicle Fuel: Used as a gasoline additive (E10, E85) or as pure fuel (E100) for modified internal combustion engines.

Chemical industry: Used as a solvent and raw material in the production of chemicals.

Electricity generation: Used in combustion turbines to produce electricity.

Ecological benefits of Bioethanol

Emissions reduction: Bioethanol reduces emissions of CO2, carbon monoxide and volatile organic compounds (VOCs).

Renewability: Bioethanol is produced from renewable raw materials, which contributes to the sustainability of energy resources.

Residue utilization: Bioethanol production from lignocellulosic biomass valorizes agricultural and forestry residues, thereby reducing waste and its environmental impact.

Economic Perspectives and Sustainable Development

The production of biodiesel and bioethanol offers promising economic prospects and contributes to sustainable development. Here are some of the main economic prospects and sustainability benefits:

Job creation: Biofuel production creates jobs in rural areas, from the cultivation of raw materials to the processing and distribution of finished products.

Diversification of agricultural income: Biofuels offer a new source of income for farmers, enabling them to diversify their activities and stabilize their incomes.

Investment Attraction: The biofuels sector attracts investment in infrastructure, technology and R&D projects. This stimulates local and regional economic development.

Support for the Circular Economy: Biofuel production supports the circular economy by recovering agricultural residues and waste, thereby reducing pressure on natural resources and minimizing waste.

Contributions to the Sustainable Development Goals (SDGs): The production and use of biofuels contributes to a number of SDGs, including those relating to clean and affordable energy, decent work and economic growth, combating climate change and sustainable resource management.

Practical Tips for Successful Biofuel Production

To succeed in biodiesel and bioethanol production, it's important to follow certain best practices and focus on key aspects. Here are a few tips to maximize your chances of success:

Selecting the right feedstocks: Choose feedstocks suited to your region and infrastructure. For example, corn and sugarcane are ideal for bioethanol, while soy and rapeseed are suitable for biodiesel.

Optimize Production Processes: Invest in modern, efficient technologies to optimize biofuel production processes. This includes the use of specific enzymes for saccharification, efficient catalysts for transesterification and advanced technologies for distillation and dehydration.

Comply with environmental standards: Make sure you meet all environmental standards when producing biofuels. Regulatory compliance is crucial to protecting the environment and maintaining consumer confidence.

Adopt Sustainable Practices: Use sustainable production practices that minimize environmental impact. This includes recycling production by-products, using renewable raw materials and responsibly managing water and soil resources.

Market research: Before launching biofuel production, carry out market research to understand market demand and trends. This will help you develop products that meet consumer expectations and identify business opportunities.

Constantly innovate: Be open to innovation and constantly seek to improve your production processes and develop new products.

Innovation is essential to staying competitive and responding to market trends.

Collaborate with Experts: Work with experts in biotechnology, engineering and agriculture to improve product quality and optimize production processes. Partnerships with research institutes and universities can also be beneficial.

Focus on Quality: Quality must be a top priority in biofuel production. High-quality products are essential for building customer loyalty and a good reputation.

Product diversification: Diversify your product range to cater for different market segments. Offer biofuels for different applications, such as biodiesel for diesel engines, bioethanol for gasoline engines and production residues for electricity generation.

Use Modern Technologies: Invest in modern technologies to improve the efficiency and quality of biofuel production. Modern equipment enables more precise and efficient processes.

Successful Case Studies

To illustrate the possibilities offered by biodiesel and bioethanol production, let's look at a few notable success stories:

Biodiesel Plant Bavaria, Germany: This plant uses local rapeseed oils to produce high-quality biodiesel. By investing in advanced transesterification technologies and complying with strict environmental standards, the plant has succeeded in optimizing biodiesel production while reducing greenhouse gas emissions. The biodiesel produced is used in local diesel vehicles and exported to international markets.

São Paulo Bioethanol Project, Brazil: This project uses sugarcane to produce high-quality bioethanol. Using state-of-the-art fermentation and distillation technologies, the project has succeeded in optimizing bioethanol production while reducing greenhouse gas emissions. The bioethanol produced is used as a gasoline additive, helping to reduce dependence on fossil fuels.

Saskatchewan Biofuels Cooperative, Canada: This cooperative transforms local canola oils into high-quality biodiesel. Using efficient and environmentally-friendly transesterification methods, the cooperative has succeeded in producing biodiesel used in local diesel engines and exported to international markets. The project has improved the incomes of member farmers and contributed to the energy sustainability of the region.

These examples show how biodiesel and bioethanol production can be an effective way of improving farm incomes, creating high-quality products and promoting sustainable development. By adopting good practices and taking advantage of market opportunities, farmers and processors can transform their raw materials into biofuels in a profitable and sustainable way.

Chapter 13: The Starch Transformation

Starch, a complex carbohydrate found in many crops such as corn, potatoes, rice and manioc, is an essential raw material for a variety of industrial products. Starch processing creates a multitude of products used in both the food and non-food sectors. This chapter explores starch processing methods, industrial applications and the economic and ecological benefits of this transformation.

Importance of starch processing

Starch is a renewable and versatile resource that offers a number of economic and environmental advantages. Here's why starch processing is important:

Added value: Starch processing adds value to agricultural crops, increasing income for farmers and processing industries.

Product diversification: Starch-derived products are used in a variety of sectors, from food to chemicals, textiles and pharmaceuticals.

Supporting sustainability: Using starch as a renewable raw material contributes to environmental sustainability by reducing dependence on fossil resources.

Innovative applications: Starch processing enables the development of innovative products, such as bioplastics, that meet sustainability and performance requirements.

Starch Processing Methods

The transformation of starch into industrial products involves several chemical and physical processes. Here are the main starch processing methods:

Enzymatic hydrolysis

Enzymatic hydrolysis is a common method of converting starch into simple sugars such as glucose, fructose and maltose. This process uses specific enzymes to break down the starch's glycosidic bonds. The main steps are as follows:

Pretreatment: Raw starch is dissolved in water to form a suspension.

Liquefied hydrolysis: alpha-amylase enzymes are added to liquefy starch, transforming it into dextrins.

Saccharification: Glucoamylase or isomerase enzymes are added to convert dextrins into simple sugars such as glucose and fructose.

Purification: Hydrolyzed sugars are purified by filtration and evaporation to obtain high-purity syrups.

Physical and chemical modification

Starch can be physically or chemically modified to improve its functional properties. Here are some common methods:

Gelatinization: Starch is heated in the presence of water, causing the starch granules to break up and release starch molecules into solution. This process is used to improve starch solubility and viscosity.

Reaction with acids or bases: Starch is treated with acids or bases to create modified starches, such as oxidized or acetylated starch. These modifications change the starch's properties, such as freeze-thaw stability and resistance to retrogradation.

Etherification and Esterification: Starch is reacted with etherifying or esterifying agents to produce starch derivatives, such as hydroxypropyl starch and starch phosphate. These derivatives have unique properties, such as enhanced water-binding capacity and improved heat resistance.

Thermal degradation

Thermal degradation is a starch processing method that uses heat to break down starch molecules into simpler compounds. This process is commonly used to produce resistant starches and pyrolysis products. Here are the main stages:

Thermal treatment: Starch is heated to high temperatures (generally between 150 and 200°C) in the absence of water. This treatment causes thermal decomposition of the starch.

Formation of Degradation Products: Thermal degradation produces simpler compounds, such as dextrins and maltodextrins, which have functional properties different from those of native starch.

Fermentation

Fermentation is a biological method of transforming starch into fermented products such as ethanol, lactic acid and amino acids. Here are the main stages:

Saccharification: Starch is hydrolyzed into fermentable sugars by enzymes.

Fermentation: Fermentable sugars are converted into fermented products by micro-organisms such as yeast and bacteria. For example, yeast converts glucose into ethanol and carbon dioxide.

Product recovery: Fermented products are separated and purified to obtain the final products, such as pure ethanol and lactic acid.

Processed starch applications

Processed starch is used in a variety of industrial sectors, each with specific applications. Here are just a few examples of processed starch applications:

Food Sector

Thickeners and gelling agents: Modified starches are used as thickeners and gelling agents in soups, sauces, desserts and dairy products. They improve the texture and stability of foods.

Glucose syrup: Glucose syrup, produced by the enzymatic hydrolysis of starch, is used as a sweetener in confectionery, beverages and bakery products.

Resistant starches: Resistant starches, produced by thermal degradation or chemical modification, are used as dietary fiber to improve digestive health and control blood sugar levels.

Non-Food Sector

Bioplastics: Modified starches are used to make bioplastics, biodegradable materials used in packaging, bags and disposable utensils. Starch-based bioplastics help reduce plastic waste.

Adhesives: Modified starches are used as binders in industrial adhesives, paper and cardboard glues, and particleboard adhesives. They offer strong adhesion and are environmentally friendly.

Pharmaceuticals: Starch is used as an excipient in tablets and capsules to improve consistency, dissolution and release of active ingredients. Starch derivatives, such as cyclodextrins, are used to increase drug solubility and stability.

Textiles: Modified starches are used in the textile industry as sizing agents to improve yarn strength during weaving. They are also used as finishers to improve fabric texture and durability.

Economic and ecological benefits

Converting starch into industrial products offers a number of economic and ecological benefits:

Creating added value: Starch processing adds value to agricultural crops, increasing income for farmers and processing industries.

Rural development: Starch production and processing can stimulate economic development in rural areas by creating jobs and improving local infrastructure.

Supporting sustainability: Using starch as a renewable raw material contributes to environmental sustainability by reducing dependence on fossil resources and minimizing waste.

Waste reduction: Starch processing enables agricultural residues and waste to be recycled, thereby reducing environmental impact and promoting a circular economy.

Practical Tips for Successful Starch Processing

To succeed in transforming starch into industrial products, it's important to follow certain best practices and focus on key aspects. Here are a few tips to maximize your chances of success:

Selecting the right raw materials: Choose starch-rich crops suited to your region and infrastructure. For example, corn and potatoes are ideal for starch production.

Optimize Production Processes: Invest in modern, efficient technologies to optimize starch transformation processes. This includes the use of specific enzymes for hydrolysis, chemical reactors for modification and fermenters for the production of fermented products.

Respect Quality Standards: Make sure you comply with all quality standards when producing starch-derived products. Regulatory compliance is crucial to ensuring consumer safety and customer confidence.

Adopt Sustainable Practices: Use sustainable production practices that minimize environmental impact. This includes recycling production by-products, using renewable raw materials and responsibly managing water and soil resources.

Market research: Before launching production of new starch-derived products, carry out market research to understand market demand and trends. This will help you develop products that meet consumer expectations and identify business opportunities.

Constantly innovate: Be open to innovation and constantly seek to improve your production processes and develop new products. Innovation is essential to staying competitive and responding to market trends.

Collaborate with Experts: Work with experts in chemistry, biotechnology and engineering to improve product quality and optimize your production processes. Partnerships with research institutes and universities can also be beneficial.

Focus on Quality: Quality must be a top priority in the production of starch-derived products. High-quality products are essential for building customer loyalty and a good reputation.

Product diversification: Diversify your product range to meet the needs of different market segments. Offer starch-derived products for different applications, such as food thickeners, bioplastics and industrial adhesives.

Use modern technologies: Invest in modern technologies to improve production efficiency and quality. Modern equipment enables more precise and efficient processes.

Successful Case Studies

To illustrate the possibilities offered by starch processing, let's look at a few notable success stories:

Corn Starch Processing Plant in Illinois, USA: This plant uses local corn to produce a range of starch-derived products, including glucose syrups, modified starches and bioplastics. By investing in advanced processing technologies and meeting strict quality standards, the plant has succeeded in optimizing production while reducing greenhouse gas emissions. The plant's by-products are used in the food and industrial sectors worldwide.

The Cassava Processing Project in Nigeria: This community-based project processes local cassava into starch and by-products such as cassava starch and bioplastics. By using sustainable processing methods and investing in modern equipment, the project has succeeded in creating jobs and generating additional income for local farmers. By-products are sold on local and international markets, contributing to the region's economic development.

La Coopérative de Transformation d'Amidon de Pomme de Terre en France: This cooperative brings together farmers specialized in potato cultivation. It transforms potatoes into high-quality starch, used in both the food and non-food sectors. By using environmentally-friendly processing methods and complying with strict quality standards, the cooperative has succeeded in increasing the added value of potato production and diversifying the income of member farmers.

These examples show how the transformation of starch into industrial products can be an effective way of improving farm incomes, creating high-quality products and promoting sustainable development. By adopting good practices and taking advantage of market opportunities, farmers and processors can transform their raw materials into starch-derived products in a profitable and sustainable way.

Chapter 14: Enzymatic Hydrolysis Transformation

Enzymatic hydrolysis is an efficient and environmentally-friendly method for transforming agricultural materials into a variety of chemical products. Using specific enzymes, this process breaks down complex biopolymers into simpler compounds, which can then be used in a variety of industrial applications. This chapter explores the techniques of enzymatic hydrolysis, the applications of this technology and the benefits it offers.

Importance of Enzymatic Hydrolysis

Enzymatic hydrolysis offers several advantages that make it the technology of choice for processing agricultural materials:

Efficiency and specificity: Enzymes are highly efficient and specific biological catalysts, enabling rapid and precise hydrolysis of substrates.

Moderate reaction conditions: Unlike traditional chemical processes, enzymatic hydrolysis takes place at moderate temperatures and pressures, reducing energy costs and minimizing environmental impact.

Safety and ecology: Enzymes are generally non-toxic and biodegradable, making the process safer for workers and less polluting for the environment.

Product quality: The use of enzymes produces high-purity products with specific properties, meeting the requirements of the food, pharmaceutical and chemical industries.

Enzyme Hydrolysis Techniques

Enzymatic hydrolysis involves the use of specific enzymes to break down complex biopolymers such as starch, proteins and lipids into simpler compounds. Here are the main steps in the process:

Enzyme selection

Amylases: Used to hydrolyze starch into simple sugars such as glucose and maltose.

Proteases: used to hydrolyze proteins into peptides and amino acids.

Lipases: Used to hydrolyze lipids into fatty acids and glycerol.

Cellulases: Used to hydrolyze cellulose into glucose.

Raw materials pre-treatment

Cleaning and grinding: Agricultural materials (grains, tubers, lignocellulosic residues) are cleaned to remove impurities and ground to increase the contact surface with the enzymes.

Hydration: Ground materials are mixed with water to form a suspension, facilitating enzyme action.

Enzymatic hydrolysis

Enzyme addition: Specific enzymes are added to the raw material suspension. The concentration and type of enzymes depend on the substrate to be hydrolyzed and the desired end product.

Reaction conditions: The enzymatic reaction takes place at moderate temperatures (30-60°C) and at a pH optimum for the enzyme used. Reaction times vary according to enzyme and substrate, from a few minutes to several hours.

Monitoring and control: Reaction parameters (temperature, pH, enzyme concentration) are monitored and adjusted to ensure efficient and complete hydrolysis.

Product Separation and Purification

Filtration: The hydrolyzed suspension is filtered to separate hydrolyzed products from solid residues.

Evaporation and drying: filtered products are concentrated by evaporation and dried to obtain high-purity powders or syrups.

Additional purification: If necessary, additional purification steps, such as chromatography or crystallization, are used to obtain products of even higher purity.

Enzyme Hydrolysis applications

Enzymatic hydrolysis has many applications in various industrial sectors. Here are just a few examples of common applications:

Food Sector

Production of Glucose and Fructose Syrup: Amylase is used to hydrolyze starch into glucose, which can then be isomerized into fructose to produce glucose-fructose syrup, commonly used as a sweetener in beverages and confectionery.

Hydrolyzed proteins: Proteases are used to produce protein hydrolysates, which are used as functional ingredients in baby foods, protein supplements and dietary products.

Lactose hydrolysis: Lactases are used to hydrolyze lactose into glucose and galactose, producing lactose-free dairy products for lactose-intolerant people.

Pharmaceuticals Sector

Amino Acid Production: Proteases hydrolyze proteins into individual amino acids, used as active ingredients in drugs and nutritional supplements.

Digestive enzymes: Digestive enzymes, such as amylases, proteases and lipases, are produced by enzymatic hydrolysis and used to treat digestive disorders.

Drug synthesis: Enzymes are used as catalysts in the synthesis of many drugs, improving the efficiency of chemical reactions and reducing unwanted by-products.

Industrial Sector

Biofuel production: Cellulases hydrolyze the cellulose in lignocellulosic residues into glucose, which is then fermented into ethanol to produce biofuels.

Detergent manufacturing: Enzymes, such as proteases and lipases, are used in detergent formulations to improve cleaning efficiency and reduce energy and water consumption.

Paper industry: Cellulases and hemicellulases are used to bleach and modify cellulose fibers, improving paper quality and reducing the use of aggressive chemicals.

Advantages of Enzyme Technology

The use of enzymes to transform agricultural materials offers several significant advantages:

Energy efficiency: Enzymatic reactions take place at moderate temperatures and pressures, reducing energy consumption compared with traditional chemical processes.

Specificity and Selectivity: Enzymes are highly specific, enabling them to precisely target the chemical bonds to be hydrolyzed, reducing unwanted by-products and improving the purity of end products.

Waste reduction: Enzymes are biodegradable and enzymatic processes generate less chemical waste, thus reducing environmental impact.

Safety and security: Enzymes are generally non-toxic and safe to handle, improving worker safety and reducing health risks.

Product quality: The products obtained by enzymatic hydrolysis are of the highest quality, with specific functional properties tailored to the needs of the food, pharmaceutical and chemical industries.

Practical Tips for Successful Enzymatic Transformation

To succeed in using enzymes to process agricultural materials, it's important to follow certain best practices and focus on key aspects. Here are a few tips to maximize your chances of success:

Selecting the right enzymes: Choose specific enzymes for your substrate and end product. For example, use amylases for starch hydrolysis and proteases for protein hydrolysis.

Optimize Reaction Conditions: Monitor and adjust enzymatic reaction parameters, such as temperature, pH and enzyme concentration, to ensure efficient and complete hydrolysis.

Use Quality Raw Materials: Ensure that agricultural raw materials are of high quality and free from contaminants that could inhibit enzymatic activity.

Invest in Modern Technologies: Use modern, efficient equipment for raw material pre-treatment, enzymatic reaction and purification of hydrolysed products.

Comply with Quality Standards: Follow strict quality standards to ensure that hydrolyzed products meet the requirements of the food, pharmaceutical and chemical industries.

Innovate and adapt: Be open to innovation and adapt your processes in line with new discoveries and technological advances. Ongoing research is essential to improve processes and develop new products.

Collaborate with Experts: Work with experts in biotechnology and enzyme chemistry to optimize your processes and guarantee product quality. Partnerships with research institutes and universities can also be beneficial.

Focus on Sustainability: Adopt sustainable production practices that minimize environmental impact and maximize resource efficiency. This includes responsible waste management and the use of renewable energy sources.

Market research: Before launching production of new hydrolyzed products, carry out market research to understand market demand and trends. This will help you develop products that meet consumer expectations and identify business opportunities.

Successful Case Studies

To illustrate the possibilities offered by enzymatic hydrolysis, let's look at a few case studies of notable successes:

The Bioethanol Plant in Iowa, USA: This plant uses cellulases to hydrolyze the cellulose in corn residues into glucose, which is then fermented into ethanol. Using high-performance enzymes and advanced fermentation technologies, the plant has succeeded in increasing bioethanol production while reducing greenhouse gas emissions. The bioethanol produced is used as a fuel for vehicles, helping to reduce dependence on fossil fuels.

The Hydrolyzed Protein Production Project in Germany: This project uses proteases to hydrolyze soy proteins into peptides and amino acids, used as functional ingredients in food supplements and dietary products. By optimizing enzymatic reaction conditions and investing in advanced purification technologies, the project has succeeded in producing high-quality hydrolyzed proteins that meet the requirements of the food and pharmaceutical industries.

The Bioplastics Manufacturing Cooperative in Brazil: This cooperative uses amylases to hydrolyze cassava starch into glucose, which is then polymerized to produce biodegradable bioplastics. By using sustainable processing methods and respecting environmental standards, the cooperative has succeeded in creating local jobs and generating additional income for farmers. The bioplastics produced are used in packaging and disposable products, helping to reduce plastic waste.

These examples show how enzymatic hydrolysis can be an effective way of improving farm incomes, creating high-quality products and promoting sustainable development. By adopting the right practices and taking advantage of market opportunities, farmers and processors can transform their raw materials into chemical products in a profitable and sustainable way.

Chapter 15: Transformation into polymerization

The conversion of agricultural materials by polymerization to produce bioplastics and other innovative materials represents a major advance in the field of green chemistry and sustainability. This process enables renewable resources to be converted into high value-added products, meeting the growing needs of industry and consumers for alternatives to traditional plastics. This chapter explores the polymerization processes, innovations and applications of bioplastics and other materials derived from agricultural materials.

Importance of the Polymerization of Agricultural Materials

The polymerization of agricultural materials offers a number of economic and ecological advantages:

Reducing dependence on fossil fuels: Bioplastics and other materials derived from the polymerization of agricultural materials reduce dependence on fossil fuels, thus contributing to energy sustainability.

Biodegradability: Many bioplastics are biodegradable, which reduces the environmental impact of plastic waste and facilitates its management.

Innovation and added value: The transformation of agricultural materials into polymeric materials enables the creation of innovative products with high added value, offering economic opportunities for farmers and processing industries.

Reduced greenhouse gas emissions: The production of bioplastics from agricultural materials generally emits less CO_2 than traditional plastics, helping to combat climate change.

Polymerization process

Polymerization is a chemical process that converts monomers into polymers. For agricultural materials, this process can include pre-treatment, fermentation and polymerization stages. Here are the main stages in the process:

Raw materials pre-treatment

Cleaning and grinding: Agricultural materials, such as corn, cassava or lignocellulosic residues, are cleaned to remove impurities and ground to increase the contact surface.

Enzymatic hydrolysis: complex biopolymers such as starch and cellulose are hydrolyzed into simple sugars by specific enzymes (amylases, cellulases). These sugars serve as substrates for fermentation.

Fermentation

Fermentation: Hydrolyzed sugars are fermented by micro-organisms to produce specific monomers, such as lactic acid, succinic acid and polyhydroxybutyrate (PHB).

Monomer recovery: The monomers produced by fermentation are recovered and purified for use in the polymerization process.

Polymerization

Condensation polymerization: This process involves the reaction of monomers with the elimination of small molecules such as water. For example, lactic acid can be polymerized to produce polylactic acid (PLA).

Addition polymerization: In this process, monomers bond without eliminating by-products. For example, PHB is produced by direct polymerization of 3-hydroxybutyrate monomers.

Co-polymerization: Different monomers can be copolymerized to create materials with specific properties tailored to particular applications.

Innovations in the Polymerization of Agricultural Materials

Innovations in the polymerization of agricultural materials have led to the development of materials with improved properties and new applications. Here are some of the most important innovations:

Nanocomposites: Incorporating nanoparticles into bioplastics improves their mechanical, thermal and barrier properties. For example, the addition of clay nanoparticles can increase the strength and rigidity of PLA.

Functionalized bioplastics: The development of bioplastics with specific functionalities, such as electrical conductivity, antimicrobial and photodegradability, expands possible applications in electronics, packaging and medicine.

Biocompatible polymers: Research into biocompatible polymers has led to innovations in medical devices, implants and matrices for controlled drug release. For example, copolymers of lactic acid and glycolic acid (PLGA) are used for absorbable sutures and drug delivery systems.

Recyclability and compostability: Improving the recyclability and compostability properties of bioplastics such as PLA and PHB facilitates their end-of-life management and reduces their environmental impact.

Applications of Bioplastics and Other Polymer Materials

Bioplastics and other polymeric materials derived from agricultural materials have a wide range of applications in various industrial sectors. Here are just a few examples of common applications:

Food and Packaging Sector

Food packaging: Bioplastics, such as PLA and PHB, are used to manufacture biodegradable food packaging, including films, containers and disposable cutlery. These materials reduce the environmental impact of packaging waste.

Bags and sacks: Bioplastics are used to produce compostable bags and sacks, offering an environmentally-friendly alternative to traditional plastic bags.

Vacuum trays and packaging: Bioplastics with improved barrier properties are used for trays and vacuum packaging, extending the shelf life of food products.

Medical Sector

Medical devices: Biocompatible bioplastics, such as PLGA, are used to manufacture medical devices, including absorbable sutures, orthopedic implants and matrices for tissue regeneration.

Drug Delivery Systems: Bioplastics are used to create controlled-release drug matrices, improving treatment efficacy and reducing side effects.

Prosthetics and Orthotics: Lightweight, durable bioplastics are used to manufacture prosthetics and orthotics, offering patients comfort and durability.

Agricultural sector

Mulching films: Biodegradable mulching films based on bioplastics such as PLA are used to improve weed management and reduce soil erosion. These films decompose naturally, eliminating the need to remove them after harvest.

Planting cups and pots: Bioplastics are used to make biodegradable planting cups and pots, making it easier to transplant plants without disturbing the roots.

Fertilizer Release Devices: Bioplastics are used to create slow-release fertilizer devices, improving fertilization efficiency and reducing nutrient losses.

Automotive and Construction sector

Automotive components: Bioplastics reinforced with natural fibers are used to manufacture lightweight, high-strength automotive components, reducing vehicle weight and improving fuel efficiency.

Building materials: Bioplastics are used to produce building materials such as insulation panels, flooring and roofing materials. These materials offer improved performance and reduce the carbon footprint of buildings.

Economic and Ecological Benefits of Bioplastics

The production of bioplastics and other polymeric materials from agricultural materials offers a number of economic and ecological advantages:

Creating added value: Processing agricultural materials into bioplastics adds value to crops, increasing income for farmers and processing industries.

Rural development: Bioplastics production can stimulate economic development in rural areas by creating jobs and improving local infrastructure.

Reducing plastic waste: Biodegradable and compostable bioplastics reduce the accumulation of plastic waste and make it easier to manage at the end of its life.

Reduced Environmental Impact: The production of bioplastics generally emits less CO_2 and consumes less energy than traditional plastics, thus helping to combat climate change.

Support for the Circular Economy: Bioplastics contribute to a circular economy by recovering agricultural residues and reducing dependence on fossil resources.

Practical Tips for Successful Bioplastics Production

To succeed in the production of bioplastics and other polymeric materials from agricultural materials, it's important to follow certain best practices and focus on key aspects. Here are a few tips to maximize your chances of success:

Selecting the right raw materials: Choose starch-, cellulose- or other biopolymer-rich crops suited to your region and infrastructure. For example, corn and manioc are ideal for starch production.

Optimize Production Processes: Invest in modern, efficient technologies to optimize pre-treatment, fermentation and polymerization processes. This includes the use of specific enzymes for hydrolysis, fermentation reactors for monomer production and polymerization reactors for polymer production.

Meet Quality Standards: Follow strict quality standards to ensure that bioplastics meet the requirements of the food, pharmaceutical and chemical industries. Regulatory compliance is crucial to ensuring consumer safety and customer confidence.

Adopt Sustainable Practices: Use sustainable production practices that minimize environmental impact. This includes recycling production by-products, using renewable raw materials and responsibly managing water and soil resources.

Market research: Before launching production of new bioplastics, carry out market research to understand market demand and trends. This will help you develop products that meet consumer expectations and identify business opportunities.

Constantly innovate: Be open to innovation and constantly seek to improve your production processes and develop new bioplastics with specific properties. Innovation is essential to remain competitive and respond to market trends.

Collaborate with Experts: Work with experts in chemistry, biotechnology and engineering to improve product quality and optimize your production processes. Partnerships with research institutes and universities can also be beneficial.

Focus on Quality: Quality must be a top priority in bioplastics production. High-quality products are essential for building customer loyalty and a good reputation.

Product diversification: Diversify your range of bioplastics to meet the needs of different market segments. Offer bioplastics for different applications, such as packaging, medical devices and construction materials.

Use modern technologies: Invest in modern technologies to improve production efficiency and quality. Modern equipment enables more precise and efficient processes.

Successful Case Studies

To illustrate the possibilities offered by the polymerization of agricultural materials, let's take a look at some notable success stories:

The Bioplastics Plant in Nebraska, USA: This plant uses local corn to produce PLA, a biodegradable bioplastic used in food packaging and disposable utensils. By investing in advanced polymerization technologies and complying with strict environmental standards, the plant has succeeded in optimizing production while reducing greenhouse gas emissions. PLA products are sold on local and international markets, helping to reduce plastic waste.

The PHB Production Project in Brazil: This project uses sugarcane residues to produce PHB, a biodegradable bioplastic used in medical devices and agricultural applications. By using sustainable fermentation methods and investing in modern equipment, the project has succeeded in creating local jobs and generating additional income for farmers. PHB products are used in hospitals and on farms, reducing the environmental impact of traditional plastics.

The Bioplastics Cooperative in Thailand: This cooperative processes cassava into bioplastics, used to manufacture biodegradable mulching films and planting pots. By using environmentally-friendly processing methods and complying with quality standards, the cooperative has succeeded in increasing the added value of cassava production and diversifying the income of its farmer-members. The bioplastics produced are sold on local and international markets, contributing to the sustainability of agriculture.

These examples show how polymerizing agricultural materials to produce bioplastics and other materials can be an effective way of improving farm incomes, creating high-quality products and promoting sustainable development. By adopting good practices and taking advantage of market opportunities, farmers and processors can transform their raw materials into bioplastics in a profitable and sustainable way.

Chapter 16: Transformation into Trans-Esterification

Transesterification is a key chemical process for the production of biodiesel from vegetable oils and animal fats. This process, which converts triglycerides into methyl esters (biodiesel) and glycerol, is essential for the creation of sustainable, renewable biofuels. This chapter explores trans-esterification techniques, the chemical processes involved, the equipment required and the applications of this technology.

The importance of trans-esterification

Transesterification has several advantages that make it a crucial process for the production of biodiesel and other chemicals:

Reducing dependence on fossil fuels: Producing biodiesel from renewable resources contributes to energy security and environmental sustainability.

Reduced greenhouse gas emissions: Biodiesel produced by transesterification emits less CO2 and other pollutants than fossil fuels.

Adding value to renewable raw materials: Processing vegetable oils and animal fats into biodiesel adds value to these raw materials, boosting the incomes of farmers and processing industries.

Production of valuable co-products: Glycerol, the co-product of transesterification, is used in a variety of industries, including cosmetics, pharmaceuticals and food.

Trans-esterification process

Transesterification is a chemical reaction that converts triglycerides (the main components of oils and fats) into methyl esters (biodiesel) and glycerol. Here are the main steps in the process:

Raw materials pre-treatment

Oils and fats used for biodiesel production need to be pre-treated to remove impurities and optimize the trans-esterification reaction. This includes filtration to remove solids, neutralization of free fatty acids, and sometimes drying to reduce water content.

Trans-esterification reaction

Addition of Reagents: Oils or fats are mixed with an alcohol (usually methanol) and a catalyst (often sodium or potassium hydroxide).

Chemical reaction: The mixture is heated to a controlled temperature (approx. 60°C) and stirred to allow the trans-esterification reaction to take place. During this reaction, triglycerides are converted into methyl esters (biodiesel) and glycerol.

Phase separation: After the reaction, the mixture is left to stand to allow the phases to separate. The less dense biodiesel floats on top of the glycerol.

Purification and Refining

Washing: Raw biodiesel is washed with water to remove catalyst, methanol and soap residues. Raw glycerol can also be purified to remove impurities and catalyst residues.

Drying: Washed biodiesel is dried to remove all residual water, while glycerol is also dried and sometimes distilled to achieve higher purity.

Final filtration: The biodiesel is filtered one last time to guarantee its purity before being stored or used.

Trans-esterification techniques and equipment

The production of biodiesel by trans-esterification requires specific equipment and the respect of certain process conditions to guarantee an efficient and safe reaction. Here are the main equipment and techniques used

Transesterification reactors

Reactors are at the heart of the transesterification process. They can be of different types, including batch and continuous reactors. Batch reactors are often used for small and medium-sized production runs, while continuous reactors are preferred for large-scale installations because of their efficiency and ability to handle large volumes of raw materials.

Batch reactors: These are closed vessels where reagents are added, mixed and heated for the reaction. They allow precise control of reaction conditions, but require downtime between batches for cleaning and preparation.

Continuous reactors: These reactors enable continuous addition of reagents and continuous recovery of finished products. They are more efficient for large production runs, and make better use of catalysts and reagents.

Mixing and Agitation Systems

Mixing and agitation are essential to ensure effective contact between the reactants (oil, methanol and catalyst) and to maintain a uniform temperature in the reactor. Mixing systems can include mechanical agitators, recirculation pumps or static mixers.

Heating systems

The trans-esterification reaction requires a controlled temperature, usually around 60°C. The heating systems used can include water heaters, heating coils or steam-heated reactors.

Separation systems

After the reaction, the product mixture needs to be separated into biodiesel and glycerol. This is usually achieved by decantation or centrifugation. Gravity decanters are commonly used for small plants, while centrifuges are preferred for large-scale production due to their efficiency and speed.

Washing and drying equipment

Raw biodiesel must be washed to remove catalyst, methanol and soap residues. Washing systems can include water washers, washing columns or dry washing systems using adsorbents. The washed biodiesel is then dried to remove any residual water, often using vacuum dryers or drying columns.

Filtration Systems

Final filtration of biodiesel is essential to guarantee its purity before use. Filtration systems can include cartridge filters, sand filters or membrane filters.

Biodiesel applications and benefits

Biodiesel produced by trans-esterification has many applications and offers a number of ecological and economic benefits.

Biodiesel applications

Transport: Biodiesel can be used as a fuel for diesel-powered vehicles. It can be used pure (B100) or blended with fossil diesel (B20, B5) to improve performance and reduce emissions.

Heating: Biodiesel can be used as fuel for domestic and industrial heating boilers, offering a renewable alternative to fossil fuels.

Electricity generation: Biodiesel can be used in diesel generators to produce electricity, particularly in rural and remote areas.

Ecological and economic benefits

Emissions reduction: Biodiesel produces less CO_2, fine particles, sulfur and polycyclic aromatic hydrocarbons (PAHs) than fossil diesel, helping to combat air pollution and climate change.

Biodegradability: Biodiesel is biodegradable and non-toxic, reducing the risk of pollution in the event of a spill and facilitating end-of-life management.

Waste recovery: The production of biodiesel from waste oils and animal fats enables us to recover this waste and reduce its environmental impact.

Local economic development: Biodiesel production stimulates economic development in rural areas by creating jobs and generating additional income for farmers and processors.

Practical Tips for Successful Biodiesel Production

To succeed in biodiesel production by trans-esterification, it's important to follow certain best practices and focus on key aspects. Here are a few tips to maximize your chances of success:

Selecting the right raw materials: Choose high-quality vegetable oils and animal fats suited to your infrastructure. For example, soybean oil, rapeseed oil and sunflower oil are ideal for biodiesel production.

Optimize Reaction Conditions: Monitor and adjust transesterification reaction parameters such as temperature, pH, catalyst concentration and oil/methanol ratio, to ensure maximum conversion of triglycerides into biodiesel.

Use efficient catalysts: Alkaline catalysts, such as sodium hydroxide and potassium hydroxide, are commonly used for transesterification. Make sure you use high-quality catalysts and dose them correctly to optimize the reaction.

Invest in modern equipment: Use modern, efficient reactors, mixing systems, separation systems and washing and drying equipment to optimize the biodiesel production process.

Meet Quality Standards: Follow strict quality standards to ensure that the biodiesel produced meets the requirements of the transportation, heating and power generation industries. Regulatory compliance is crucial to ensure consumer safety and customer confidence.

Adopt Sustainable Practices: Use sustainable production practices that minimize environmental impact. This includes recycling production by-products, using renewable raw materials and responsibly managing water and soil resources.

Market research: Before launching biodiesel production, carry out market research to understand market demand and trends. This will help you develop products that meet consumer expectations and identify business opportunities.

Constantly innovate: Be open to innovation and constantly seek to improve your production processes and develop new products. Innovation is essential to staying competitive and responding to market trends.

Collaborate with Experts: Work with experts in chemistry, engineering and biotechnology to improve product quality and optimize your production processes. Partnerships with research institutes and universities can also be beneficial.

Focus on Quality: Quality must be a top priority in biodiesel production. High-quality products are essential for building customer loyalty and a good reputation.

Successful Case Studies

To illustrate the possibilities offered by trans-esterification, let's look at a few case studies of notable successes:

Argentina Biodiesel Plant: This plant uses local soybean oil to produce high-quality biodiesel. By investing in advanced trans-esterification technologies and complying with strict environmental standards, the plant has succeeded in optimizing production while reducing greenhouse gas emissions. The biodiesel produced is used in local diesel vehicles and exported to international markets.

The Biodiesel Project in India: This project uses used cooking oil to produce biodiesel, reducing waste and valorizing an otherwise neglected resource. By using sustainable trans-esterification methods and investing in modern equipment, the project has succeeded in creating local jobs and generating additional income for farmers. The biodiesel produced is used in public transport, reducing the environmental impact of the sector

The Biodiesel Cooperative in Kenya: This cooperative transforms animal fats into biodiesel, used for electricity generators and industrial heating boilers. By using environmentally-friendly trans-esterification methods and complying with quality standards, the cooperative has succeeded in increasing the added value of animal fat production and diversifying the income of member farmers. The biodiesel produced is sold on local and international markets, contributing to the region's energy sustainability.

These examples show how trans-esterification for biodiesel production can be an effective way of improving farm incomes, creating high-quality products and promoting sustainable development. By adopting the right practices and taking advantage of market opportunities, farmers and processors can transform their raw materials into biodiesel in a profitable and sustainable way.

Chapter 17: Transformation into Green Chemistry

Green chemistry, also known as sustainable chemistry, is an approach that aims to design chemical products and processes that reduce or eliminate the use and generation of hazardous substances. When applied to the processing of agricultural products, green chemistry promotes more environmentally-friendly practices, supports sustainability and reduces negative impacts on human health and the ecosystem. This chapter explores the principles of green chemistry, their application to the processing of agricultural products, and the benefits in terms of sustainability and respect for the environment.

Principles of Green Chemistry

Green chemistry is based on twelve fundamental principles that guide chemists in designing safer and more efficient processes and products:

Prevention: Avoiding the formation of waste is more effective than treating or cleaning it up after it has been produced.

Atom Economy: Designing syntheses to maximize the incorporation of all materials used in the final product.

Safe Synthesis: Use substances and synthesis methods that reduce or eliminate toxicity to humans and the environment.

Safe Product Design: Producing chemicals that perform their function while being less toxic.

Solvents and safe auxiliaries: Minimize the use of auxiliary substances (such as solvents) and, if necessary, use safe substances.

Energy efficiency: conduct chemical reactions at ambient temperatures and pressures to save energy.

Use of Renewable Materials: Favoring the use of renewable raw materials rather than non-renewable resources.

Derivative reduction: Avoid using blocked or protected derivatives whenever possible, as these steps require additional reagents and generate waste.

Catalysis: Using catalysts rather than stoichiometric reagents to increase the efficiency of chemical reactions.

Design for Degradation: Designing chemical products that break down into harmless substances after use.

Real-Time Analysis for Pollution Prevention: Monitor chemical processes in real time to prevent the formation of hazardous substances.

Safe Inherent Chemistry: Choosing substances and forms of chemical substances to minimize the risk of chemical accidents.

Applications of Green Chemistry to the Processing of Agricultural Products

Green chemistry can be applied to various aspects of agricultural product processing, from the production of biofuels and bioplastics to the synthesis of pharmaceuticals and cosmetics.

Biofuel production: The application of green chemistry in biofuel production involves the use of biological catalysts (enzymes) to hydrolyze biopolymers such as starch and cellulose into simple sugars, which are then fermented to produce ethanol or biogas. The use of enzymes reduces the need for harsh reaction conditions and toxic chemicals.

Bioplastics manufacturing: Green chemistry is used to polymerize monomers derived from agricultural raw materials, such as lactic acid obtained from sugar fermentation, into polylactic acid (PLA). This process uses safe solvents and catalysts, and the PLA produced is biodegradable.

Essential oil extraction: Green chemistry techniques for essential oil extraction include the use of supercritical CO2, a method that uses carbon dioxide under high pressure as a solvent. This technique is safer and more environmentally friendly than traditional methods using organic solvents.

Synthesis of Pharmaceuticals: Green chemistry is applied to synthesize active pharmaceutical ingredients from medicinal plants. For example, alkaloids can be extracted and modified using enzyme-catalyzed reactions, reducing the use of toxic reagents and the waste produced.

Cosmetics manufacturing: the principles of green chemistry are used to produce natural and organic cosmetics. Vegetable oils and plant extracts are transformed into creams, lotions and other cosmetics using gentle methods and safe solvents.

Green Chemistry Techniques and Equipment

To apply the principles of green chemistry to the processing of agricultural products, several specific techniques and equipment can be used:

Biocatalysts: Enzymes are efficient biological catalysts that can be used to accelerate chemical reactions at moderate temperatures and pressures. For example, amylases can be used to hydrolyze starch into glucose, and lipases to trans-esterify oils into biodiesel.

Continuous Flow Reactors: Continuous flow reactors enable better utilization of reagents and catalysts, and can be designed to minimize waste and maximize energy efficiency. They are particularly useful for reactions requiring precise conditions and real-time control.

Supercritical CO2 extraction: This technique uses high-pressure carbon dioxide as a solvent to extract bioactive compounds from plants. It is more environmentally friendly than extraction methods using volatile organic solvents.

Photocatalytic reactors: These reactors use light to activate catalysts that accelerate chemical reactions. They are used for processes such as the degradation of pollutants or the synthesis of chemicals from renewable raw materials.

Real-time analysis: Real-time analysis systems monitor chemical reactions and adjust reaction conditions to minimize waste and prevent the formation of hazardous substances.

Sustainability and respect for the environment

Applying green chemistry principles to the processing of agricultural products contributes to sustainability and respect for the environment in several ways:

Reduced use of fossil resources: By using renewable raw materials, such as agricultural crops, to produce chemicals and materials, green chemistry reduces dependence on non-renewable fossil resources.

Reducing waste and emissions: Green chemistry processes are designed to minimize waste and pollutant emissions, thus contributing to the protection of air, water and soil.

Improved energy efficiency: By using chemical reactions that take place at ambient temperatures and pressures, green chemistry reduces energy consumption and associated costs.

Promoting biodegradability and recyclability: Chemicals and materials produced by green chemistry methods are often designed to be biodegradable or recyclable, facilitating their end-of-life management and reducing their environmental impact.

Health and safety: By avoiding the use of toxic substances and minimizing the risk of chemical accidents, green chemistry protects the health of workers and consumers alike.

Practical Tips for Success in Green Chemistry

To succeed in applying the principles of green chemistry to the processing of agricultural products, it's important to follow certain best practices and focus on key aspects. Here are a few tips to maximize your chances of success:

Selecting Renewable Raw Materials: Use agricultural crops and biomass residues as raw materials for the production of chemicals and materials. Choose crop varieties suited to your region and growing conditions.

Optimize Reaction Conditions: Monitor and adjust chemical reaction parameters, such as temperature, pH and reagent concentration, to maximize efficiency and minimize waste.

Use Safe Catalysts and Solvents: Use enzymatic catalysts and environmentally-friendly solvents, such as water and supercritical CO_2, to reduce the toxicity and environmental impact of chemical reactions.

Invest in modern equipment: Use continuous-flow reactors, real-time analysis systems and other modern equipment to optimize production processes and guarantee product quality.

Comply with Quality and Safety Standards: Follow strict quality and safety standards to ensure that chemicals and materials meet the requirements of the food, pharmaceutical and chemical industries. Regulatory compliance is crucial to ensuring consumer safety and customer confidence.

Adopt Sustainable Practices: Use sustainable production practices that minimize environmental impact. This includes recycling production by-products, using renewable raw materials and responsibly managing water and soil resources.

Innovate and adapt: Be open to innovation and adapt your processes in line with new discoveries and technological advances. Ongoing research is essential to improve processes and develop new products.

Collaborate with Experts: Work with experts in chemistry, biotechnology and engineering to improve product quality and optimize your production processes. Partnerships with research institutes and universities can also be beneficial.

Focus on Quality: Quality must be a top priority in the production of chemicals and materials derived from agricultural processing. High-quality products are essential for building customer loyalty and a good reputation.

Product diversification: Diversify your product range to cater for different market segments. Offer chemicals and materials for different applications, such as bioplastics, cosmetics and pharmaceuticals.

Use modern technologies: Invest in modern technologies to improve production efficiency and quality. Modern equipment enables more precise and efficient processes.

Successful Case Studies

To illustrate the possibilities offered by green chemistry applied to the processing of agricultural products, let's take a look at a few notable success stories:

The Bioplastics Plant in Nebraska, USA: This plant uses local corn to produce PLA, a biodegradable bioplastic used in food packaging and disposable utensils. By investing in advanced polymerization technologies and complying with strict environmental standards, the plant has succeeded in optimizing production while reducing greenhouse gas emissions. PLA products are sold on local and international markets, helping to reduce plastic waste.

The PHB Production Project in Brazil: This project uses sugarcane residues to produce PHB, a biodegradable bioplastic used in medical devices and agricultural applications. By using sustainable fermentation methods and investing in modern equipment, the project has succeeded in creating local jobs and generating additional income for farmers. PHB products are used in hospitals and on farms, reducing the environmental impact of traditional plastics.

The Bioplastics Cooperative in Thailand: This cooperative processes cassava into bioplastics, used to manufacture biodegradable mulching film and planting pots. By using environmentally-friendly processing methods and complying with quality standards, the cooperative has succeeded in increasing the added value of cassava production and diversifying the income of its farmer-members. The bioplastics produced are sold on local and international markets, contributing to the sustainability of agriculture.

These examples show how green chemistry can be applied to the processing of agricultural products to improve farm incomes, create high-quality products and promote sustainable development. By adopting good practices and taking advantage of market opportunities, farmers and processors can transform their raw materials into chemicals and materials in a profitable and sustainable way.

Chapter 18: Fiber conversion

The production of natural fibers from agricultural materials is a promising approach to creating sustainable, environmentally-friendly materials for textile and industrial applications. This transformation adds value to agricultural resources by producing biodegradable and renewable fibers that meet the growing needs of the textile industry and other industrial sectors. This chapter explores natural fiber production techniques, market prospects and the economic and ecological benefits associated with this transformation.

Importance of Fiber Transformation

Processing agricultural materials into natural fibres offers several significant advantages:

Sustainability: Natural fibers are renewable and biodegradable, reducing environmental impact compared to petroleum-derived synthetic fibers.

Adding value to agricultural resources: This transformation enables agricultural by-products and residues to be put to good use, boosting farmers' and producers' incomes.

Diversified applications: Natural fibers are used in a variety of applications, including textiles, composites, insulation materials and geotextiles.

Reducing dependence on synthetic fibers: Using natural fibers helps reduce dependence on synthetic fibers and fossil resources, promoting a more circular economy.

Natural Fiber Production Techniques

The production of natural fibers from agricultural materials involves several stages, from plant cultivation to fiber processing. Here are the main natural fiber production techniques:

Fibre plant cultivation

Fibrous plants are grown specifically for fiber production. The main crops include :

Cotton: Grown for its fibers used in the manufacture of textiles.

Flax: Used to produce flax fibers, which are transformed into fabrics and industrial products.

Hemp: Used to produce hemp fibers, which are transformed into textiles, ropes and composite materials.

Jute: Used to produce jute fibers, mainly used in the manufacture of bags and carpets.

Ramie: A fibrous plant producing fine, strong fibers used in high-quality textiles.

Harvesting and preparation

Fibrous plants are harvested at the optimum time to maximize fiber quality and quantity. After harvesting, the plants are prepared for fiber extraction by processes such as retting, threshing and combing.

Retting: This process involves fermenting plant stems in water or on soil to break down the pectin and separate the fibers from the stems.

Threshing: The retted stems are threshed to remove the woody parts and free the fibers.

Combing: Raw fibers are combed to align the fibers and remove impurities.

Fiber Extraction

Fibres are extracted from plants using mechanical or chemical methods:

Mechanical extraction: The fibers are extracted by machines that break the stalks and release the fibers. This method is commonly used for flax and hemp.

Chemical extraction: Fibres are extracted using chemicals to dissolve the non-fibrous parts of the plant. This method is used for certain fibers such as ramie.

Fiber Processing

The raw fibres extracted are transformed into yarns and fabrics by spinning, weaving and knitting processes:

Spinning: Fibers are spun into yarn using spinning machines. Spinning can be dry or wet, depending on the type of fiber.

Weaving: Yarns are woven into fabrics using looms. Weaving can produce a variety of fabric structures, including flat fabrics, twills and satins.

Knitting: Yarns are knitted into textiles using knitting machines. Knitting is commonly used to produce knitwear, jerseys and stitches.

Treatment and Finishing

Fabrics and products made from natural fibres undergo treatments and finishes to improve their properties and appearance:

Bleaching : Fabrics can be bleached to remove impurities and improve color.

Dyeing: Fabrics are dyed to add color and pattern.

Finishing treatments: Fabrics can be given finishing treatments to improve strength, softness, waterproofing and other specific properties.

Natural fiber applications

Natural fibers produced from agricultural materials have a wide range of applications in various industrial sectors:

Textile sector

Clothing: Natural fibers such as cotton, linen and hemp are used to make comfortable, durable clothing.

Household linen: Natural fibers are used to produce sheets, towels, curtains and other household linen.

Technical fabrics: Natural fibers are used to manufacture technical fabrics for specific applications, such as protective clothing, medical textiles and geotextiles.

Industrial Sector

Composites: Natural fibers are used as reinforcements in composite materials to produce strong, lightweight parts for the automotive, aeronautics and construction industries.

Insulation materials: Natural fibers are used to manufacture thermal and acoustic insulation materials for buildings.

Ropes and bags: Natural fibers like hemp and jute are used to make ropes, bags, nets and other industrial products.

Agricultural sector

Geotextiles: Natural fibers are used to manufacture biodegradable geotextiles for soil stabilization, erosion prevention and crop protection.

Mulch: Natural fibers are used to produce biodegradable mulch, improving soil water retention and reducing weeds.

Health Sector

Medical products: Natural fibers are used to make dressings, sutures and other medical products.

Hygiene: Natural fibers are used to produce hygiene products such as sanitary towels, diapers and wipes.

Market outlook for natural fibers

The market for natural fibers is growing rapidly, driven by increasing demand for sustainable, environmentally-friendly products. Here are some market prospects for natural fibers:

Growing demand for eco-friendly products: Consumers are increasingly aware of the environmental impact of the products they buy, and are looking for sustainable alternatives. Natural fibers meet this demand by offering a renewable and biodegradable option to synthetic fibers.

Innovation and New Product Development: Advances in natural fiber processing and treatment technologies enable the development of new products with improved properties. For example, natural fiber-reinforced composites offer equivalent or superior performance to synthetic composites.

Political and regulatory support: Many governments and international organizations are encouraging the use of sustainable materials and the reduction of plastic waste. This creates a favorable environment for the development and adoption of natural fibers.

Market opportunities in emerging countries: Emerging countries, where agricultural resources are abundant, present significant opportunities for the production and processing of natural fibers. The development of this industry can stimulate economic growth and create jobs in rural areas.

Growth in the Technical Textiles Market: Natural fibers are finding more and more applications in technical textiles, where their specific properties, such as strength, lightness and biodegradability, are particularly sought-after. This includes applications in protective clothing, geotextiles and building materials.

Practical Tips for Successful Natural Fiber Production

To succeed in the production of natural fibers from agricultural materials, it's important to follow certain best practices and focus on key aspects. Here are a few tips to maximize your chances of success:

Select Suitable Crops: Choose fibre plants that are adapted to your region and growing conditions. Be sure to select varieties that offer good fiber quality and high yields.

Optimize Growing Practices: Use optimized growing practices to maximize fiber quality and quantity. This includes soil management, irrigation, fertilization and protection against pests and diseases.

Invest in Modern Equipment: Use modern equipment for harvesting, retting, threshing, combing and fiber processing. Modern machinery improves production efficiency and quality.

Comply with Quality Standards: Follow strict quality standards to ensure that the fibers produced meet the requirements of the textile and industrial industries. Regulatory compliance is crucial to ensuring consumer safety and customer confidence.

Adopt Sustainable Practices: Use sustainable production practices that minimize environmental impact. This includes using organic growing techniques, managing water and soil resources responsibly, and recycling production by-products.

Innovate and adapt: Be open to innovation and adapt your processes in line with new discoveries and technological advances. Ongoing research is essential to improve processes and develop new products.

Collaborate with Experts: Work with experts in agriculture, biotechnology and engineering to improve product quality and optimize production processes. Partnerships with research institutes and universities can also be beneficial.

Product diversification: Diversify your product range to meet the needs of different market segments. Offer natural fibers for different applications, such as textiles, composites and insulation materials.

Use modern technologies: Invest in modern technologies to improve production efficiency and quality. Modern equipment enables more precise and efficient processes.

Successful Case Studies

To illustrate the possibilities offered by the transformation of agricultural materials into natural fibres, let's look at a few notable success stories:

L'Usine de Fibres de Chanvre in France: This plant uses local hemp to produce high-quality fibers used in textiles, composites and building materials. By investing in modern processing technologies and complying with strict environmental standards, the plant has succeeded in optimizing production while reducing greenhouse gas emissions. Hemp fiber products are sold on local and international markets, contributing to the reduction of plastic waste and the sustainability of the textile industry.

The Jute Fiber Project in India: This project uses jute residues to produce natural fibers used in the manufacture of bags, carpets and geotextiles. By using sustainable processing methods and investing in modern equipment, the project has succeeded in creating local jobs and generating additional income for farmers. Jute fiber products are used in the agricultural and construction industries, reducing the environmental impact of synthetic materials.

La Coopérative de Fibres de Lin in Belgium: This cooperative processes flax into natural fibers used to manufacture high-quality textiles, composite materials and insulation products. By using environmentally-friendly processing methods and complying with quality standards, the cooperative has succeeded in increasing the added value of flax production and diversifying the income of member farmers. The flax fibers produced are sold on local and international markets, contributing to the sustainability of the textile industry and the reduction of plastic waste.

These examples show how converting agricultural materials into natural fibers can be an effective way of improving farm incomes, creating high-quality products and promoting sustainable development. By adopting good practices and taking advantage of market opportunities, farmers and processors can transform their raw materials into natural fibers in a profitable and sustainable way.

Chapter 19: Traditional agriculture

Traditional agriculture, practiced for millennia, is the foundation of many rural societies around the world. It is based on ancestral methods and techniques handed down from generation to generation. This type of farming is deeply rooted in cultural traditions and has significant economic implications. This chapter explores traditional farming methods, their cultural significance and the associated economic implications.

Description of Traditional Farming Methods

Traditional agriculture encompasses a variety of practices and techniques that differ from region to region and from crop to crop. Here are some of the methods commonly used in traditional agriculture:

Subsistence farming

Subsistence farming is characterized by the production of crops and livestock primarily for family consumption, with little or no surplus for sale. The main characteristics of this method include:

Diversified crops: Farmers grow a variety of crops to ensure a balanced diet and reduce the risk of crop failure.

Crop rotations: Crop rotations are used to maintain soil fertility and reduce pest infestations.

Polyculture: Farmers practice polyculture, i.e. the simultaneous cultivation of several species on the same plot of land, to maximize resource use and increase the resilience of farming systems.

Itinerant farming

Shifting cultivation, also known as slash-and-burn farming, involves clearing and burning a plot of forest to grow crops for a few years, before leaving the plot fallow to regenerate. The main features of this method include:

Clearing and burning: Farmers clear and burn vegetation to release nutrients trapped in the biomass.

Fallow periods: After a few years of cultivation, the plot is left fallow to allow natural regeneration of soil fertility.

Mobility: Farmers regularly move their crops to new plots to avoid soil exhaustion.

Agriculture en Terrasses

Terraced farming is a technique of cultivation on steep slopes, where terraces are built to create flat surfaces and prevent soil erosion. The main features of this method include:

Terrace construction: Terraces are built by excavating horizontal levels on the slopes, supported by stone or earth walls.

Water management: Terraces enable efficient water management, reducing runoff and increasing water infiltration into the soil.

Soil preservation: Terraces reduce soil erosion and maintain fertility by preventing the loss of arable land.

Traditional Irrigation Agriculture

Traditional irrigation involves ancient water management techniques for irrigating crops. The main features of this method include:

Irrigation canals: Irrigation canals are dug to bring water from rivers, lakes or springs to the fields.

Water harvesting systems: Water harvesting systems, such as reservoirs and ponds, are used to store water during the rainy season for use during the dry season.

Water regulation: Farmers use water regulation techniques, such as dikes and dams, to control the flow and distribution of water.

Traditional farming

Traditional breeding involves managing domestic animals according to ancestral methods. The main characteristics of this method include :

Extensive grazing: Animals are raised on extensive pasture, often on common land or natural rangelands.

Traditional breeds: Farmers raise local breeds of livestock adapted to environmental conditions and resistant to disease.

Breeding techniques: Traditional breeding techniques, such as natural selection and controlled mating, are used to maintain herd quality and genetic diversity.

Cultural Importance of Traditional Agriculture

Traditional agriculture is deeply rooted in the cultures and traditions of rural communities. It plays a crucial role in preserving ancestral knowledge and practices, and in maintaining social cohesion and cultural identity.

Knowledge transfer

Traditional agricultural knowledge and techniques are passed down from generation to generation through informal learning, rituals and ceremonies. This transmission of knowledge contributes to the preservation of cultural practices and the resilience of rural communities.

Rituals and Ceremonies

Traditional farming is often accompanied by rituals and ceremonies that mark important stages in the agricultural cycle, such as sowing, harvesting and harvest festivals. These rituals strengthen community ties and celebrate farmers' cultural heritage.

Value and Belief Systems

The value and belief systems of rural communities are closely linked to their farming practices. Traditional agriculture reflects values of respect for nature, sustainability and cooperation. It also incorporates spiritual and religious beliefs that influence farming practices and natural resource management.

Economic Implications of Traditional Agriculture

Traditional agriculture has significant economic implications for rural communities. It provides livelihoods, generates income and contributes to food security. However, it also presents economic challenges that need to be addressed.

Food Safety

Traditional agriculture plays a crucial role in the food security of rural communities. By producing a diversity of crops and animals, it ensures a balanced diet and reduces dependence on external markets. In addition, polyculture and crop rotation techniques contribute to the resilience of farming systems to climatic hazards and pests.

Revenue generation

Although traditional agriculture is mainly oriented towards self-consumption, it can also generate income by selling surplus production on local markets. Traditional agricultural products, such as fruit, vegetables, cereals, dairy products and meats, are often highly prized for their quality and craftsmanship.

Economic challenges

Traditional agriculture presents a number of economic challenges, including :

Market access: Traditional farmers may find it difficult to access markets due to geographical isolation, lack of infrastructure and poor integration into value chains.

Access to credit: The lack of guarantees and financial resources limits traditional farmers' access to credit and the financing they need to invest in improving their farming practices.

Competitiveness: Traditional agricultural products can be less competitive on the market due to low productivity and higher production costs compared to modern farming methods.

Solutions and Opportunities

To overcome these economic challenges, several solutions and opportunities can be envisaged:

Organization and cooperation: Traditional farmers can organize themselves into cooperatives or associations to improve their access to markets, share resources and knowledge, and strengthen their bargaining power.

Adding value to products: Certifying traditional agricultural products as organic, fair trade or geographical origin products can increase their market value and attract consumers concerned with quality and sustainability.

Training and innovation: Training traditional farmers in modern, sustainable techniques, and adopting appropriate technologies, can improve the productivity and profitability of their farms.

Access to financing: The development of financing mechanisms tailored to the needs of traditional farmers, such as microcredit and guarantee funds, can facilitate access to credit and encourage investment in improved farming practices.

Successful Case Studies

To illustrate the possible successes of traditional agriculture, let's look at a few notable case studies:

The Subak Irrigation System in Bali, Indonesia: Subak is a traditional irrigation system used by Balinese farmers to cultivate rice in terraces. This community-based system, over a thousand years old, is based on the principles of cooperation and sustainable water management. Thanks to social organization and water regulation, Subak farmers have maintained stable and sustainable rice production, contributing to food security and the preservation of Balinese culture.

Shifting cultivation among the indigenous peoples of the Amazon: The indigenous peoples of the Amazon practice shifting cultivation, which involves the clearing and temporary cultivation of forest plots. This method, adapted to the environmental conditions of the rainforest, maintains soil fertility and preserves biodiversity. Indigenous communities use traditional ecological knowledge to manage natural resources sustainably, ensuring their subsistence while conserving the ecosystem.

Jardins de Case in Senegal: Jardins de Case, or family gardens, are plots of land located around rural dwellings in Senegal. These gardens, cultivated mainly by women, produce a wide variety of vegetables, fruit, herbs and medicinal plants. Home gardens contribute to food security, nutrition and women's empowerment, while preserving traditional agricultural knowledge and practices.

These examples show how traditional agriculture can be an effective way of improving food security, generating income and promoting sustainable development. By adopting good practices and overcoming economic challenges, traditional farmers can leverage their ancestral knowledge to create resilient and sustainable farming systems.

Traditional agriculture remains an important pillar of rural societies, combining ancestral know-how and sustainable practices to meet modern challenges. It offers a valuable alternative to intensive farming methods, focusing on the preservation of natural resources, crop diversity and social cohesion. For farmers and communities wishing to adopt a more sustainable and environmentally-friendly approach, traditional agriculture represents a viable and rewarding option.

Chapter 20: Intensive agriculture

Intensive agriculture is a farming method that aims to maximize production on small surfaces by using advanced technologies, chemical inputs and rigorous management practices. Developed mainly during the 20th century, this approach has transformed agriculture by dramatically increasing productivity. However, it is also the subject of debate due to its environmental and social impacts. This chapter explores intensive farming practices, their advantages and disadvantages in terms of productivity and environmental impact.

Intensive farming practices

Intensive agriculture is based on a series of practices and technologies designed to maximize agricultural production. Here are some of the main practices:

Use of Chemical Inputs: Intensive agriculture makes extensive use of chemical fertilizers to improve soil fertility and increase crop yields. Pesticides and herbicides are also used to control pests and weeds, ensuring optimum production.

Modernized irrigation: advanced irrigation systems, such as drip irrigation and center pivot irrigation, are used to deliver water to crops in an efficient, targeted way. This maximizes water use and minimizes losses.

Agricultural machinery: The use of modern agricultural machinery, such as tractors, combines and sprayers, makes it possible to mechanize many farming tasks, increasing productivity and reducing the need for labor.

Selection and Genetic Engineering: Plant varieties and animal breeds are selected and genetically improved to increase their yield, disease resistance and tolerance to adverse environmental conditions.

Integrated Crop Management: Farmers practice integrated crop management, which combines the use of chemical inputs, cultivation techniques and management practices to optimize production and reduce the risk of losses.

Intensive livestock farming: In intensive livestock farming, animals are often raised in confined or semi-confined systems, where their feed, reproduction and health are tightly controlled to maximize meat, milk or egg production.

Advantages of Intensive Agriculture

Intensive agriculture offers several significant advantages, particularly in terms of productivity and food security:

Increased productivity: Intensive agriculture enables the production of large quantities of crops and animal products on small areas. This increase in productivity is essential to meet the growing food demand of an expanding world population.

Economic efficiency: Mechanization and the use of chemical inputs reduce labor costs and optimize resources, improving the economic efficiency of farms.

Food availability: By increasing agricultural production, intensive farming contributes to food availability, reducing the risk of food shortages and famines.

Technological innovations: Intensive agriculture encourages technological innovation and the development of new farming techniques, which can improve the resilience and long-term sustainability of farming systems.

Specialization and economies of scale: Intensive farms can specialize in the production of specific crops or animal products, benefiting from economies of scale and productivity gains.

Disadvantages of Intensive Agriculture

Despite its advantages, intensive agriculture also has a number of disadvantages, particularly in terms of environmental impact and sustainability:

Soil degradation: Intensive use of chemical fertilizers and pesticides can lead to soil degradation, reducing its natural fertility and capacity to retain water. This degradation can lead to soil erosion and the loss of arable land.

Water pollution: Fertilizers and pesticides can contaminate water sources, leading to the pollution of rivers, lakes and groundwater. This pollution can have harmful effects on aquatic biodiversity and the quality of drinking water.

Greenhouse gas emissions: Intensive agricultural practices, such as the use of farm machinery and intensive livestock farming, contribute to greenhouse gas emissions, exacerbating climate change. The use of nitrogen fertilizers can also lead to emissions of nitrous oxide, a powerful greenhouse gas.

Biodiversity loss: Monoculture and pesticide use can reduce the biodiversity of agro-ecosystems, affecting populations of pollinating insects, birds and other beneficial species. Biodiversity loss can reduce the resilience of agricultural systems to disturbance.

Pesticide and antibiotic resistance: Intensive pesticide use can lead to the development of resistance in pests, necessitating the use of increasing quantities of chemicals. Similarly, the use of antibiotics in intensive livestock farming can lead to the emergence of antibiotic-resistant bacteria, posing risks to human health.

Animal health and welfare: In intensive livestock farming, animals are often raised in confined conditions that can affect their health and welfare. Stress, disease and abnormal behavior can be more prevalent in these systems.

Dependence on chemical inputs: Intensive agriculture creates a dependence on chemical inputs, such as fertilizers and pesticides, which can lead to high costs for farmers and negative environmental impacts.

Sustainable Practices in Intensive Agriculture

To mitigate the negative impacts of intensive farming and promote more sustainable agriculture, several practices can be implemented:

Conservation Agriculture: Conservation agriculture combines techniques such as direct seeding, crop rotation and permanent plant cover to improve soil health, reduce erosion and increase biodiversity.

Integrated Pest Management (IPM): IPM uses a combination of biological, cultural and chemical methods to control pests effectively and sustainably, reducing dependence on pesticides.

Organic and biological fertilizers: The use of compost, manure and other organic fertilizers can improve soil fertility and reduce dependence on chemical fertilizers. Green manures and cover crops can also enrich soils with nutrients.

Economic Water Resources: Adopting efficient irrigation techniques, such as drip irrigation, can reduce water consumption and minimize losses through evaporation and runoff.

Extensive Livestock and Agroforestry: Integrating extensive livestock and agroforestry can diversify farming systems, improve resilience and reduce environmental impacts. Trees can provide shade, wildlife habitat and valuable ecosystem services.

Use of Resilient Varieties and Breeds: Selecting crop varieties and animal breeds that are resilient to extreme weather conditions, disease and pests can improve the sustainability and productivity of intensive farming systems.

Precision technologies: Precision agriculture uses advanced technologies such as sensors, drones and GPS systems to optimize the use of inputs and improve the efficiency of farming practices. This reduces losses and minimizes environmental impact.

Prospects for Intensive Agriculture

The future of intensive agriculture will depend on the ability of farmers, researchers and policy-makers to integrate sustainable practices and meet environmental and social challenges. Here are some prospects for intensive agriculture:

Technological innovation: Technological advances will continue to play a key role in improving the efficiency and sustainability of intensive agriculture. Innovations in biotechnologies, precision technologies and resource management systems will help to meet the challenges of food production.

Policies and regulations: Policies and regulations will play a crucial role in promoting sustainable farming practices. Economic incentives, subsidies and environmental standards can encourage farmers to adopt more environmentally-friendly practices.

Education and training: Training farmers in sustainable techniques and providing them with access to up-to-date scientific and technical information will be essential to foster the transition to more sustainable intensive agriculture.

Partnerships and Cooperation: Cooperation between farmers, researchers, industries and governments will be essential to develop innovative and sustainable solutions. Public-private partnerships can facilitate technology transfer and the implementation of sustainable practices on a large scale.

Consumer involvement: Consumers play an important role in influencing farming practices through their purchasing choices. Growing demand for high-quality, sustainable products can encourage farmers to adopt more environmentally-friendly practices.

Adapting to climate change: Intensive agriculture will have to adapt to the impacts of climate change, such as temperature variations, droughts and floods. Adopting resilient practices and diversifying farming systems will be essential to maintain productivity and food security.

In conclusion, intensive agriculture offers significant opportunities to increase productivity and meet global food demand. However, it also presents environmental and social challenges that require appropriate attention and management. By integrating sustainable practices and adopting a balanced approach, intensive agriculture can contribute to a more resilient and sustainable food system, capable of feeding a growing world population while preserving natural resources and ecosystem health.

Chapter 21: Organic farming

Organic farming is a method of agricultural production that emphasizes the use of sustainable, environmentally-friendly practices. It avoids the use of synthetic chemicals, favoring instead natural processes to maintain soil fertility, control pests and diseases, and promote biodiversity. This chapter explores the techniques and principles of organic farming, the certification process and the market for organic products.

Principles of Organic Farming

Organic farming is based on several fundamental principles that guide farming practices and management decisions:

Soil, plant, animal and human health: Organic farming aims to promote the health and well-being of soils, plants, animals and humans, using practices that reinforce natural biological cycles and biodiversity.

Ecology: Organic practices are designed to work in harmony with natural ecosystems, using local, renewable resources and minimizing environmental impact.

Fairness: Organic farming is committed to treating all stakeholders fairly, including farmers, farm workers, consumers and local communities, by promoting fair and transparent relationships.

Precaution: Decisions in organic farming are taken with caution, assessing the risks and potential impacts on the environment and human health before introducing new technologies or practices.

Organic farming techniques

Organic farmers use a variety of techniques to maintain soil fertility, control pests and diseases, and promote biodiversity. Here are some of the main techniques:

Crop rotation: Crop rotation involves alternating crop types on the same plot from one season to the next to prevent soil exhaustion and reduce pest and disease infestations.

Organic fertilizers: Organic farmers use organic fertilizers, such as compost, manure and green manure, to enrich the soil with nutrients and improve its structure.

Pest and Disease Management: Integrated Pest Management (IPM) is used to control pests and diseases by combining biological, cultural and mechanical methods. This includes the use of natural predators, traps, physical barriers and repellent plants.

Cover crops: Cover crops, such as legumes, are planted to protect the soil from erosion, improve soil structure and fix atmospheric nitrogen.

Mulching: Mulching involves covering the soil with organic materials, such as straw or wood shavings, to conserve moisture, suppress weeds and add organic matter to the soil.

Selecting resistant varieties: Organic farmers select plant varieties and animal breeds that are naturally resistant to disease and adapted to local conditions.

Water management: Organic farming uses sustainable water management techniques, such as drip irrigation, to minimize water consumption and maximize irrigation efficiency.

Biodiversity: Crop diversification and the integration of agroforestry practices are encouraged to promote biodiversity and create resilient agricultural ecosystems.

Organic Agriculture certification

Organic certification is a process that guarantees that agricultural products are grown and processed according to established organic standards. Certification is important to ensure consumer confidence and the integrity of the organic market. Here are the main steps in the certification process:

Application for Certification: Farmers and processors interested in organic certification must submit an application to an accredited certification body. The application includes information on farming practices, inputs used and farm management plans.

Inspection: An on-site inspection is carried out by a certified inspector to verify that farming practices comply with organic standards. The inspection includes an assessment of fields, processing facilities, records and management practices.

Compliance Review: The results of the inspection are examined by the certification body, which determines whether the farm or processing facility complies with organic standards.

Certification: If all requirements are met, the certification body issues an organic certificate, enabling the farmer or processor to market their products as organic.

Monitoring and Renewal: Organic certification is an ongoing process that requires annual inspections and regular monitoring to ensure that farming and processing practices remain compliant with organic standards.

Organic Products Market

The organic produce market is growing rapidly, driven by increasing consumer demand for healthy, sustainable food free from synthetic chemicals. Here are some trends and prospects for the organic products market:

Demand growth: Global demand for organic products has grown significantly over the last few decades, particularly in developed countries. Consumers are increasingly aware of the health and environmental benefits of organic products.

Product diversification: The organic market has diversified to include a wide range of products, including fresh fruit and vegetables, dairy products, meats, processed products, beverages, textiles and cosmetics.

Prices and premiums: Organic products are often sold at higher prices than conventional products, due to the additional costs associated with sustainable farming practices and certification. Organic farmers can benefit from price premiums for their products.

Distribution channels: Organic products are available through a variety of distribution channels, including supermarkets, farmers' markets, specialty stores and online sales. Short-distance distribution systems, such as direct-to-consumer sales and organic basket subscriptions, are also popular.

Supporting policies: Many governments and international organizations support the development of organic farming through subsidies, training programs, research and favorable policies. These initiatives aim to encourage the transition to sustainable farming practices and strengthen the market for organic products.

Certification and labelling: Certification and labelling of organic products play a crucial role in differentiating products on the market and guaranteeing authenticity for consumers. Organic labels are widely recognized and valued by consumers.

Advantages of organic farming

Organic farming offers many benefits for the environment, human health and farming communities:

Environmental protection: By avoiding the use of synthetic chemicals, organic farming helps protect soil, water and biodiversity. Organic practices promote soil regeneration, water conservation and the creation of habitats for wildlife.

Human health: Organic products are free of pesticide residues and chemical fertilizers, reducing health risks for consumers. What's more, organic foods are often richer in nutrients and antioxidants.

Animal welfare: Organic farming emphasizes animal welfare by providing living conditions adapted to their natural needs, such as access to the outdoors, spacious living areas and organic feed.

Resilience and Sustainability: Organic farming systems are more resilient to climatic and economic shocks, due to crop diversification, soil health and sustainable management practices. Organic farming contributes to the long-term sustainability of farms.

Farmer empowerment: Organic farming encourages farmer empowerment by reducing dependence on chemical inputs and promoting the use of local and renewable resources. Organic farmers can also benefit from premium prices for their produce.

Challenges and opportunities of organic farming

Despite its many advantages, organic farming also presents challenges that must be met to ensure its success and growth:

Yields: Yields of organic crops may be lower than those of conventional crops, particularly during the transition period to organic practices. Farmers need to adopt effective management techniques to maintain and improve yields.

Certification: The organic certification process can be costly and complex, particularly for small-scale farmers. It is important to provide technical and financial support to facilitate access to certification.

Competition: The organic market is increasingly competitive, with a growing number of producers and processors. Organic farmers need to stand out for the quality of their products and their sustainable practices.

Education and awareness: It is essential to educate consumers about the benefits of organic products, and to raise farmers' awareness of organic practices. Awareness campaigns and training programs can play a crucial role in promoting organic farming.

Research and innovation: Research and innovation are essential to develop new techniques and varieties adapted to organic farming. Partnerships between farmers, researchers and research institutions can foster innovation and the improvement of organic practices.

Market access: Facilitating organic farmers' access to markets is crucial to the sector's growth. Marketing initiatives, distribution infrastructures and support policies can help overcome barriers to market access.

Successful Case Studies

To illustrate the success of organic farming, let's look at a few notable case studies:

Rodale Organic Farm, USA: Rodale Farm is a pioneer of organic farming in the USA, demonstrating the benefits of organic practices since the 1940s. Using techniques such as crop rotation, composting and integrated pest management, the farm has succeeded in maintaining high yields and improving soil health.

The European Organic Farm Network: Across Europe, many organic farms work in networks to share knowledge, resources and innovations. These networks facilitate the transition to organic farming and improve farm resilience. For example, the BioCultural Heritage Territories project in Spain integrates traditional and organic farming practices to promote sustainability and cultural diversity.

Organic Urban Agriculture in Cuba: In Cuba, organic urban agriculture emerged in response to food shortages after the fall of the Soviet Union. Organic urban gardens, known as organopónicos, produce a large proportion of the fresh vegetables consumed in Cuban cities. These gardens use organic techniques such as composting, biological control and crop rotation to ensure sustainable production.

These examples show how organic farming can be an effective way of improving the sustainability, health and resilience of farming systems. By adopting organic practices and overcoming the associated challenges, farmers can contribute to a more sustainable and equitable agricultural future.

Chapter 22: Urban agriculture

Urban agriculture is the increasingly popular practice of growing plants and raising animals in urban and peri-urban environments. This approach responds to many contemporary challenges, such as food insecurity, lack of green space and environmental problems. Urban agriculture is developing thanks to technological innovations and community initiatives that are transforming cities into centers of sustainable food production. This chapter explores the development of urban agriculture, innovations, challenges and benefits for urban communities.

Development of Urban Agriculture

Urban agriculture encompasses a variety of practices and models, from community gardens to vertical farms. The development of urban agriculture is based on several factors:

Urban population growth: As the urban population grows, so does the demand for fresh, local food. Urban agriculture makes it possible to meet this demand by producing food directly in cities.

Limited space: The space available for traditional agriculture is limited in urban areas. Urban agriculture maximizes the use of available spaces, such as rooftops, walls, vacant lots and balconies.

Advanced technologies: Technological innovations such as hydroponics, aquaponics and vertical growing systems enable plants to be grown efficiently and productively in restricted urban environments.

Community initiatives: Community initiatives and non-profit organizations play a crucial role in the development of urban agriculture, mobilizing residents, providing resources and creating support networks.

Support policies: Municipal policies and government support programs encourage the development of urban agriculture by providing incentives, subsidies and infrastructure.

Innovations in Urban Agriculture

Urban agriculture is marked by many innovations that make food production more efficient and sustainable. Here are some of the main innovations:

Hydroponics: Hydroponics allows plants to be grown without soil, using a nutrient solution. Hydroponic systems can be installed on roofs, balconies and inside buildings, maximizing the use of space and reducing water consumption.

Aquaponics: Aquaponics combines aquaculture (fish farming) and hydroponics in a symbiotic system. Fish waste provides nutrients for the plants, while the plants filter the water for the fish. This integrated system is sustainable and highly productive.

Vertical Farming: Vertical farms use stacked structures to grow plants at height, optimizing the use of vertical space. Vertical growing technologies include growing towers, mobile shelves and plant walls.

Rooftop gardens: The roofs of buildings are transformed into gardens or farms, providing additional growing space in dense urban areas. Rooftop gardens also contribute to the thermal regulation of buildings and the management of rainwater.

Urban greenhouses: Urban greenhouses use glass or plastic structures to create controlled growing environments. They extend growing seasons and improve yields using advanced climate management techniques.

Connected Agriculture: Connected urban agriculture uses sensors, real-time monitoring systems and Internet of Things (IoT) technologies to optimize growing conditions and resource management. Farmers can monitor and adjust parameters such as humidity, temperature and nutrient levels remotely.

Challenges of Urban Agriculture

Despite its many advantages, urban agriculture presents a number of challenges that must be overcome to ensure its success and sustainability:

Access to space: Finding space available for agriculture in densely populated urban areas can be difficult. Competition for urban land use is intense, and available space can be expensive.

Initial costs: The initial costs of establishing urban agriculture systems, such as greenhouses, hydroponic systems and vertical farms, can be high. Urban farmers need to find ways of financing these investments.

Technical skills: Urban agriculture requires specific technical skills, especially for advanced systems such as hydroponics and aquaponics. Training and education are essential to ensure the success of urban farmers.

Regulations and policies: Municipal regulations can limit the development of urban agriculture. Urban farmers must navigate complex regulatory frameworks to obtain necessary permits and comply with health and safety standards.

Resource management: Efficient management of resources such as water, energy and nutrients is crucial to the sustainability of urban agriculture. Farmers need to adopt sustainable management practices to minimize environmental impact.

Community Awareness and Acceptance: Gaining the acceptance and support of urban residents is important to the success of urban agriculture projects. Initiatives must be well communicated and integrated into local communities.

Benefits for urban communities

Urban agriculture offers many benefits for urban communities, from improving food security to promoting environmental sustainability. Here are some of the main benefits:

Food security: Urban agriculture improves food security by producing fresh, nutritious food directly in cities. This reduces dependence on long and vulnerable supply chains and ensures constant access to quality food.

Reduced carbon footprint: By producing food locally, urban agriculture reduces transport distances and associated greenhouse gas emissions. It also helps reduce packaging and energy consumption.

Improved air quality: Plants grown in urban environments absorb carbon dioxide and release oxygen, helping to improve air quality. Urban gardens and green walls can also reduce air pollution levels.

Job and income creation: Urban agriculture creates employment opportunities for urban residents, particularly in food production, farming systems management and retailing. It can also generate additional income for urban farmers.

Education and awareness: Urban agriculture projects offer educational opportunities for residents, especially children and young people. They can learn about food production, sustainability and the importance of agriculture.

Community building: Community gardens and urban farms create meeting and socializing spaces for residents. They strengthen community ties and encourage cooperation and mutual aid.

Reducing food waste: Urban agriculture can help reduce food waste by using composting techniques and recycling organic waste to enrich soil and feed plants.

Improving Urban Resilience: By diversifying sources of food supply and creating local production systems, urban agriculture improves the resilience of cities to economic and environmental disruption.

Successful Case Studies

To illustrate the success of urban agriculture, let's look at a few notable case studies:

Sky Greens Vertical Farm in Singapore: Sky Greens is an innovative vertical farm located in Singapore, a city-state with little arable land. The farm uses rotating growing towers to produce fresh vegetables efficiently and sustainably. The towers are powered by a low-energy hydraulic system, and water is recycled to minimize waste. Sky Greens supplies fresh vegetables to local supermarkets, reducing dependence on food imports.

Brooklyn Grange Rooftop Gardens, New York, USA: Brooklyn Grange operates several rooftop gardens in New York City, producing vegetables, herbs and honey for local markets and restaurants. Brooklyn Grange's rooftop gardens also contribute to stormwater management and building thermal regulation. In addition to food production, Brooklyn Grange offers educational programs and community events.

Organopónicos in Havana, Cuba: In response to food shortages in the 1990s, Havana developed a network of urban gardens called organopónicos. These gardens use organic techniques and local inputs to produce a large proportion of the fresh vegetables consumed in the city. Organopónicos have improved food security, created jobs and strengthened urban resilience in Cuba.

These examples show how urban agriculture can transform cities into centers of sustainable food production, while delivering significant social, economic and environmental benefits. By adopting technological innovations and overcoming challenges, urban communities can leverage urban agriculture to create more resilient, healthy and sustainable environments.

Chapter 23: Agroforestry

Agroforestry is an agricultural practice that combines farming and forestry by integrating trees and shrubs into crop and livestock systems. This approach aims to create diversified, sustainable agricultural ecosystems that improve farm productivity, resilience and biodiversity. Agroforestry offers many benefits, both for the environment and for farmers. This chapter explores the principles of agroforestry, current techniques and the benefits for biodiversity and agricultural productivity.

Agroforestry principles

Agroforestry is based on several fundamental principles that guide the design and management of agroforestry systems:

Diversity: Agroforestry promotes the diversity of plant and animal species by integrating trees, shrubs, annual and perennial crops, and livestock into a single system. This diversity improves the resilience of farming systems to environmental and economic stresses.

Complementarity: Plant and animal species in agroforestry systems are chosen for their complementary interactions. Trees and shrubs can provide shade, nutrients, habitat and wind protection for crops and animals.

Sustainability: Agroforestry aims to improve the sustainability of agricultural systems by preserving natural resources, improving soil fertility and increasing the capacity of ecosystems to regenerate.

Circular Economy: Agroforestry systems seek to maximize the use of local resources and minimize waste by recycling nutrients and integrating sustainable management practices.

Agroforestry techniques

Agroforestry encompasses a variety of techniques and practices that can be adapted to local conditions and farmers' objectives. Here are just a few of the common agroforestry techniques:

Agroforestry Sylvopastoral Systems: These systems combine animal husbandry with tree and shrub management. The animals graze under the trees, which provide shade and complementary fodder. Trees can also be harvested for wood, firewood or fruit.

Silvoarable Agroforestry Systems: These systems integrate annual or perennial crops with trees and shrubs. Crops are planted between rows of trees, diversifying production and improving soil fertility by adding organic matter from the trees.

Hedges and windbreaks: Hedges and windbreaks are rows of trees and shrubs planted around or within fields to protect crops from high winds, reduce soil erosion and provide habitat for wildlife. Hedges can also produce fruit, nuts, firewood and other useful products.

Forest Gardens: Forest gardens mimic the structure of natural forests by integrating fruit trees, shrubs, perennials and annual crops into a stratified system. This approach maximizes the use of vertical space and creates a diverse, productive ecosystem.

Reforestation Agroforestry: This technique consists of reforesting degraded or marginal land by integrating crops and trees. Reforestation improves soil quality, prevents erosion and contributes to carbon sequestration.

Integrated Agroforestry Systems: These systems combine several agroforestry practices on a single farm to maximize ecological and economic benefits. For example, a farm might include silvopastoral systems, hedgerows, forest gardens and intercropping.

Benefits for biodiversity

Agroforestry offers many benefits for biodiversity by creating diverse habitats and improving the resilience of agricultural ecosystems:

Wildlife habitat: Trees and shrubs in agroforestry systems provide habitats for a variety of animal species, including birds, insect pollinators, mammals and reptiles. These habitats promote biodiversity and help regulate pest populations.

Improving soil quality: Trees and shrubs improve soil quality by increasing organic matter, fixing nitrogen (in the case of legumes) and improving soil structure. Better-quality soils encourage plant growth and the diversity of soil organisms.

Pollination and pest control: Plant diversity in agroforestry systems attracts a variety of pollinating insects and natural predators of pests. This improves crop pollination and reduces the need for pesticides.

Conservation of endemic species: Agroforestry conserves local plant and animal species that may be threatened by intensive agricultural practices. The genetic diversity of local species is preserved, which is essential for ecosystem resilience.

Carbon sequestration: Trees and shrubs sequester carbon in their biomass and soils, helping to combat climate change. Agroforestry systems can store more carbon than conventional agricultural systems.

Benefits for Agricultural Productivity

In addition to the benefits for biodiversity, agroforestry also improves agricultural productivity in several ways:

Improving soil fertility: Trees and shrubs enrich soils with organic matter, increasing their capacity to retain water and nutrients. This improves soil fertility and crop productivity.

Erosion protection: Agroforestry systems reduce soil erosion by stabilizing slopes and protecting soils from wind and rain. Less erosion means better conservation of fertile soils and more stable agricultural production.

Favorable microclimate: Trees create favorable microclimates by providing shade and reducing wind speed. This protects crops from extreme temperatures and adverse weather conditions, improving yields.

Income diversification: Agroforestry diversifies sources of income by producing a variety of products, such as fruit, nuts, wood, medicinal products and fodder. This diversification reduces economic risks for farmers.

Reduced input costs: By improving soil fertility and providing natural fodder and nutrients, agroforestry can reduce dependence on chemical fertilizers and pesticides. This reduces input costs for farmers.

Shock resilience: Agroforestry systems are more resilient to climatic and economic shocks due to their diversity and ability to regenerate. Farmers can better cope with droughts, floods and market price fluctuations.

Successful Case Studies

To illustrate the benefits of agroforestry, let's look at a few notable case studies:

Silvopastoral systems in France: In France, many farmers adopt silvopastoral systems, integrating fruit and forest trees with pasture for their animals. These systems improve soil fertility, provide shade for animals and produce fruit and firewood. Farmers benefit from income diversification and greater resilience to climatic hazards.

Agroforestry in India: In India, agroforestry is used to combat land degradation and improve food security. Farmers plant trees such as neem, mango and tamarind in their food crop fields. These trees provide fruit, medicine and firewood, while improving soil fertility and reducing erosion.

Forest gardens in West Africa: In West Africa, traditional forest gardens integrate a wide variety of plants, including fruit trees, medicinal plants and food crops. These agroforestry systems provide a varied diet, additional income and medicinal products, while conserving local biodiversity.

Hedges and windbreaks in the UK: In the UK, traditional hedges are used to demarcate fields.

Chapter 24: Precision farming

Precision agriculture is a modern approach to farm management that uses cutting-edge technologies to optimize production. It relies on the use of detailed data and advanced tools to make informed decisions, improve resource efficiency, increase yields and reduce environmental impacts. This chapter explores the principles of precision agriculture, the technologies used and the importance of data and precise crop management.

Principles of Precision Agriculture

Precision farming is based on several fundamental principles:

Data Collection: Precision farming involves collecting detailed data on soil conditions, crops, weather and other factors influencing agricultural production. These data are essential for understanding spatial and temporal variations within fields.

Site-specific management: Unlike conventional agriculture, which often treats fields as homogeneous entities, precision agriculture recognizes variations within fields and adapts farming practices accordingly. Each area of the field can be specifically managed to optimize production.

Input optimization: By using precise data, precision agriculture enables the use of inputs such as fertilizers, pesticides and water to be optimized. This reduces costs, minimizes environmental impact and improves the sustainability of farming systems.

Data-driven decision-making: Farming decisions are made based on data analysis and predictive models. This enables farmers to react quickly to changes and optimize farming operations.

Technologies Used in Precision Agriculture

Precision agriculture uses a variety of advanced technologies to collect and analyze data, and to implement precise farming practices. Here are some of the main technologies used:

Global Positioning Systems (GPS): GPS enables farmers to map their fields precisely and track farming operations with great accuracy. GPS-equipped agricultural equipment can be used for site-specific applications, such as planting, fertilizing and spraying.

Soil and Crop Sensors: Sensors are used to measure parameters such as soil moisture, temperature, electrical conductivity, nutrient content and crop health. These sensors provide real-time data that can be used to adjust crop management practices.

Drones and aerial imaging: Drones equipped with multispectral and thermal cameras are used to monitor crops and soils. They can capture detailed images that reveal information on plant health, pest infestations, diseases and nutrient deficiencies.

Geographical Information Systems (GIS): GIS enables geospatial data to be collected, stored, analyzed and visualized. Farmers can use GIS to map variations within fields and to plan specific management interventions.

Predictive models and Big Data: Predictive models and Big Data analyses use algorithms to analyze large quantities of data and predict future trends. These tools can help farmers anticipate water, nutrient and crop protection needs.

Intelligent Agricultural Machinery: Intelligent agricultural equipment, such as autonomous tractors and precision sprayers, are equipped with sensors and control systems that enable precise, automated operations. These machines can adjust input application rates in real time based on collected data.

The Importance of Data in Precision Agriculture

Data plays a central role in precision agriculture. The efficient collection, analysis and use of data can optimize crop management and improve farming results.

High-precision data collection: Collecting precise, detailed data is the first step in precision agriculture. Sensors, drones, satellites and GPS equipment collect data on a variety of parameters, such as soil properties, crop growth, weather conditions and nutrient levels.

Data analysis and interpretation: Collected data must be analyzed and interpreted to provide useful information to farmers. Data analysis software and predictive models are used to process the data and identify trends, anomalies and specific crop needs.

Informed Decision Making: Farming decisions are made based on information derived from data analysis. Farmers can adjust crop management practices, such as irrigation, fertilization and crop protection, according to the specific needs of each area of the field.

Resource optimization: Effective use of data enables the optimal use of agricultural resources, such as water, fertilizers and pesticides. This reduces costs, minimizes environmental impacts and improves the sustainability of farming systems.

Monitoring and evaluation: Data is also used to monitor and evaluate the performance of crops and management practices. Farmers can monitor results in real time and make adjustments if necessary to maximize yields and crop quality.

Benefits of Precision Farming

Precision farming offers numerous benefits for farmers, the environment and society in general:

Increased yields: By optimizing crop management practices, precision farming increases yields and crop quality. Farmers can produce more with fewer resources.

Cost reduction: Efficient use of inputs and resources reduces production costs. Farmers can save on fertilizers, pesticides, water and energy.

Improved sustainability: Precision farming minimizes environmental impact by reducing excessive use of chemical inputs and preserving natural resources. This contributes to the long-term sustainability of agricultural systems.

Resilience to Climate Change: Precision technologies enable farmers to adapt to climate variations in real time. They can adjust management practices according to weather conditions and climate forecasts, improving crop resilience.

Efficient Resource Management: Precision farming enables efficient management of water, nutrient and energy resources. Farmers can make more targeted and efficient use of these resources.

Reduced environmental impact: By reducing chemical inputs and optimizing management practices, precision farming helps reduce soil, water and air pollution. This protects ecosystems and biodiversity.

The challenges of Precision Agriculture

Despite its many advantages, precision farming also presents challenges that must be overcome to ensure its adoption and success:

High initial costs: Precision technologies such as sensors, drones and intelligent machinery systems can be costly to implement. Farmers need to invest in these technologies and in the training required to use them effectively.

Technical complexity: Precision farming requires advanced technical skills to collect, analyze and interpret data. Farmers need to be trained in the use of precision technologies and data analysis software.

Technology access : Access to precision technologies may be limited in some regions, particularly in developing countries. Infrastructure, connectivity and financial resources may represent obstacles to the adoption of precision agriculture.

Data integration : Collecting data from different sources and integrating it into a coherent system can be complex. Farmers need effective data management platforms to maximize the benefits of precision agriculture.

Adapting to small farms: Small farms may find it difficult to adopt precision technologies due to cost and complexity. Solutions tailored to small farms need to be developed to enable wider adoption.

Successful Case Studies

To illustrate the successes of precision agriculture, let's look at a few notable case studies:

John Deere and the Farm Machinery Revolution: John Deere, a leading farm machinery company, has developed a range of intelligent tractors and equipment equipped with precision technologies. These machines use GPS, sensors and analysis software to optimize farming operations, reducing costs and increasing yields for farmers.

Wageningen University project in the Netherlands: Wageningen University conducted a research project on precision agriculture in potato crops. Using drones, soil sensors and predictive models, researchers were able to optimize irrigation, fertilization and crop protection, increasing yields and reducing the use of chemical inputs.

Agricultural cooperative in Argentina: An agricultural cooperative in Argentina has adopted precision farming to improve the management of soybean and corn crops. Using GPS, soil sensors and analysis software, cooperative members were able to optimize management practices, reducing costs and increasing yields.

These examples show how precision agriculture can transform farming practices, improve productivity and promote sustainability. By adopting precision technologies and overcoming challenges, farmers can harness the benefits of precision agriculture to create more efficient, resilient and environmentally friendly farming systems.

Chapter 25: Subsistence farming

Subsistence farming is a form of agriculture in which farmers concentrate primarily on producing food for themselves and their families. This practice is common in many rural areas of the world, particularly in developing countries. Subsistence farming plays a crucial role in the food security of rural families, providing them with staple foods and ensuring their survival in times of crisis. This chapter explores subsistence farming practices, its role in food security and the challenges faced by subsistence farmers.

Subsistence farming practices

Subsistence farming is based on methods and techniques that enable farmers to maximize food production on small areas of land. Here are just a few of the common subsistence farming practices:

Polyculture: Polyculture is the cultivation of several plant species on the same plot of land. This technique diversifies food sources, reduces the risk of crop failure and improves soil fertility through complementary crops.

Crop rotation: Crop rotation is a practice in which farmers alternate crop types on the same plot from one season to the next. This technique helps maintain soil fertility, control pests and diseases, and reduce nutrient depletion.

Use of local varieties: Subsistence farmers often prefer local and traditional plant varieties, which are better adapted to local environmental conditions and more resistant to disease and pests.

Organic fertilizers: The use of organic fertilizers, such as compost and manure, is common in subsistence farming. These fertilizers improve soil fertility and provide plants with the nutrients they need in a sustainable way.

Water conservation techniques: Subsistence farmers use water conservation techniques such as terraces, bunds and reservoirs to maximize water use and ensure efficient crop irrigation.

Agroforestry: Integrating trees and shrubs into farming systems is a common practice in subsistence agriculture. Trees provide shade, nutrients, additional food products and enhance biodiversity.

Animal husbandry: Raising domestic animals, such as chickens, goats and cattle, is often associated with subsistence farming. Animals provide sources of protein, milk, eggs and manure to fertilize the soil.

Fallow land: Fallow land means leaving a parcel of land at rest for a period to allow the soil's fertility to regenerate naturally. This technique is often used in rotation with crops.

The role of subsistence agriculture in food security

Subsistence farming plays an essential role in the food security of rural families, providing them with staple foods and reducing their dependence on external markets. Here's how subsistence farming contributes to food security:

Staple food production: Subsistence farmers produce staple foods such as cereals, vegetables, fruit and legumes, which are the main source of nutrition for their families. The diversity of crops grown ensures a balanced diet.

Reduced dependence on markets: By producing their own food, subsistence families reduce their dependence on local markets for their food needs. This is particularly important in times of economic crisis or fluctuating food prices.

Shock resilience: Subsistence farming offers a degree of resilience to economic and climatic shocks. Subsistence farming families can better cope with market disruptions and extreme weather conditions by having direct access to their own food resources.

Use of Local Resources: Subsistence farming relies on the use of local resources, such as traditional plant varieties, organic fertilizers and water conservation techniques. This helps maintain the sustainability of farming systems and preserve local ecosystems.

Cost savings: By producing their own food, subsistence families save on the cost of purchasing foodstuffs. This is particularly important for low-income families, who can allocate their limited resources to other essential needs.

Challenges of Subsistence Farming

Despite its many advantages, subsistence farming also presents several challenges that can affect the food security of rural families:

Limited Access to Resources: Subsistence farmers often have limited access to resources such as land, water, improved seeds and agricultural inputs. This can reduce their ability to produce enough food to meet their needs.

Low productivity: Traditional subsistence farming techniques can result in low crop and animal productivity. Limited yields may not be sufficient to feed families, particularly in times of drought or crop failure.

Vulnerability to Climate Shocks: Subsistence farmers are particularly vulnerable to climate shocks such as droughts, floods and storms. These events can destroy crops and infrastructure, jeopardizing families' food security.

Limited access to markets and services: Subsistence farmers often have limited access to markets to sell surplus production and purchase agricultural inputs. They may also lack support services such as training, credit and transport infrastructure.

Limited knowledge and skills: Subsistence farmers may have limited knowledge and skills in modern farming techniques, resource management and sustainable practices. This can affect their ability to improve the productivity and resilience of their farming systems.

Health and nutrition: Although subsistence farming provides a food base, it may not always meet families' complete nutritional needs. Diets may lack diversity and essential nutrients.

Solutions and Opportunities for Subsistence Farming

To overcome the challenges of subsistence farming and improve food security for rural families, several solutions and opportunities can be envisaged:

Training and Education: Providing training and education to subsistence farmers on modern farming techniques, sustainable practices and resource management can improve their productivity and resilience. Agricultural extension programs and advisory services can play a crucial role in this respect.

Access to inputs and resources: Facilitating subsistence farmers' access to agricultural inputs, such as improved seeds, fertilizers and tools, as well as natural resources, such as water and land, can improve their ability to produce enough food.

Local market development: Promoting the development of local markets and transport infrastructures can help subsistence farmers to sell their surplus production and access the necessary inputs. Agricultural cooperatives and producer associations can facilitate access to markets.

Climate risk management: Setting up climate risk management systems, such as crop insurance, early warning systems and climate adaptation techniques, can help subsistence farmers cope with climate shocks and reduce their vulnerability.

Diversification of crops and sources of income: Encouraging diversification of crops and sources of income can improve the food security and economic resilience of subsistence families. Agroforestry practices, animal husbandry and non-agricultural activities can contribute to this diversification.

Enhancing nutritional security: Promoting nutritional security by encouraging dietary diversity and the production of nutrient-rich crops can improve the health and well-being of subsistence families. Vegetable gardens and nutrition programs can play an important role.

Access to credit and finance: Facilitating access to credit and finance for subsistence farmers can enable them to invest in improving their farming practices and infrastructure. Microcredit and financing programs tailored to the needs of small-scale farmers can be particularly useful.

Successful Case Studies

To illustrate the success of subsistence farming, let's look at a few notable case studies:

Kitchen garden program in Zimbabwe: In Zimbabwe, a kitchen garden program has been set up to help subsistence families improve their food security and nutrition. Families receive seeds, tools and training in gardening techniques. The vegetable gardens provide a regular source of fresh, varied vegetables, improving dietary diversity and family health.

Agroforestry project in Kenya: In Kenya, an agroforestry project has been launched to help subsistence farmers improve the productivity of their land while preserving natural resources. Farmers plant fruit trees and nitrogen-fixing trees among their food crops. The trees improve soil fertility, provide additional food products and increase farmers' incomes.

Agricultural Training Initiative in India: In India, an agricultural training initiative has been set up to teach subsistence farmers modern, sustainable techniques. Farmers learn to use organic fertilizers, manage water efficiently and diversify their crops. The training improves crop yields, reduces production costs and strengthens family food security.

These examples show how targeted initiatives can improve the food security, productivity and resilience of subsistence farmers. By adopting modern, sustainable farming practices, facilitating access to resources and building farmers' capacities, it is possible to transform subsistence farming and ensure a more secure and prosperous future for rural families.

Chapter 26: Conservation Agriculture

Conservation agriculture is a sustainable approach to farmland management that aims to preserve natural resources, particularly soil and water, while maintaining or improving agricultural productivity. This method is based on a set of farming practices designed to minimize soil disturbance, protect vegetation cover and improve soil health. This chapter explores soil and natural resource conservation methods in agriculture, as well as the associated techniques and ecological benefits.

Soil Conservation Methods

Soil conservation is at the heart of conservation agriculture. The following practices are essential for maintaining soil fertility, preventing erosion and improving soil structure:

No-till and reduced tillage: No-till, or direct seeding, and reduced tillage are practices that minimize soil disturbance. By avoiding or reducing tillage, soil structure is preserved, organic matter is protected, and erosion is reduced. These practices also help maintain soil moisture and promote soil biodiversity.

Permanent Plant Cover: Maintaining a permanent vegetation cover, whether by cover crops, crop residues or perennial plants, protects the soil from erosion, reduces water runoff and improves soil fertility. Cover crops, such as legumes, fix nitrogen in the soil and increase organic matter.

Crop rotation: Crop rotation is the practice of alternating crop types on the same plot from one season to the next. This technique reduces nutrient depletion, improves soil structure, and helps control pests and diseases. Crop diversity also promotes biodiversity.

Agroforestry and hedgerows: Integrating trees, shrubs and hedges into farming systems protects soil from erosion, improves water retention and provides habitats for wildlife. The roots of trees and shrubs stabilize the soil, while their foliage reduces the impact of rainfall on the soil.

Terraces: Terraces are stepped structures built on slopes to slow water runoff and reduce erosion. They also make it possible to cultivate sloping areas, preserving topsoil and improving water retention.

Mulching: Mulching involves covering the soil with organic materials such as straw, wood shavings or dead leaves. Mulching reduces water evaporation, suppresses weeds and adds organic matter to the soil as it decomposes.

Natural Resource Conservation Methods

In addition to soil conservation, conservation agriculture focuses on the sustainable management of other natural resources, including water, nutrients and biodiversity:

Efficient water management: Efficient water management is crucial to conservation agriculture. Efficient irrigation techniques, such as drip and sprinkler irrigation, reduce water consumption and minimize losses through evaporation and runoff. Rainwater harvesting and the construction of reservoirs and bunds help store water for periods of drought.

Organic amendments: The use of organic amendments, such as compost and manure, improves soil fertility, increases organic matter and strengthens the soil's capacity to retain water and nutrients. These amendments also promote soil biological activity and biodiversity.

Integrated Nutrient Management: Integrated nutrient management combines the use of organic and mineral fertilizers to meet crop needs while minimizing nutrient losses through leaching and volatilization. Regular soil analyses help determine specific nutrient requirements and adjust applications accordingly.

Functional Biodiversity: The promotion of functional biodiversity, i.e. the diversity of species that provide ecosystem services, is essential in conservation agriculture. Natural predators, pollinators and soil micro-organisms play key roles in pest regulation, crop pollination and organic matter decomposition.

Buffer zones and riparian buffer strips: Buffer zones and riparian buffer strips are areas of natural or planted vegetation located along watercourses and bodies of water. They filter pollutants, reduce runoff and erosion, and provide habitats for wildlife.

Conservation Agriculture Techniques

The following techniques are commonly used in conservation agriculture to achieve sustainability and productivity goals:

Direct seeding and no-till : Direct seeding involves sowing seeds directly into crop residues without tilling the soil. This technique reduces soil disturbance, preserves soil structure and increases organic matter. Non-tillage is similar, but also involves the use of plant cover to protect the soil.

Cover crops: Cover crops, such as legumes, grasses and crucifers, are sown between main crops to protect the soil and improve its fertility. They fix nitrogen, add organic matter and reduce weeds.

Crop rotation: Crop rotation involves alternating crops on the same plot. It reduces the risk of disease and pests, improves soil structure, and diversifies farmers' sources of income.

Agroforestry: Agroforestry integrates trees and shrubs into agricultural systems. Trees provide shade, nutrients and enhance biodiversity. They can also provide products such as fruit, wood and nuts.

Integrated Pest Management (IPM): IPM combines biological, cultural and chemical methods to control pests in a sustainable way. Natural predators, resistant crops and habitat management techniques are used to minimize pesticide use.

Organic amendments: The application of compost, manure and other organic matter improves soil structure, increases water retention capacity, and provides essential nutrients for crops. Organic amendments also promote biological activity in the soil.

Ecological benefits of Conservation Agriculture

Conservation agriculture offers numerous ecological benefits that contribute to the sustainability of agricultural systems and environmental protection:

Soil erosion prevention: Soil conservation practices such as no-till and cover crops reduce erosion caused by wind and water. Preserving topsoil improves long-term productivity and protects soil resources.

Improved Soil Quality: Adding organic matter and improving soil structure increase fertility, water retention capacity and biological activity. Healthy soils support productive and resilient crops.

Water Conservation: Effective irrigation techniques and water management practices reduce water consumption and increase water use efficiency. This is particularly important in drought-prone regions.

Carbon sequestration: Conservation agriculture contributes to carbon sequestration by increasing soil organic matter and integrating trees and shrubs into farming systems. This helps mitigate climate change.

Promoting Biodiversity: Conservation practices promote biodiversity by creating habitats for wildlife, reducing pesticide use, and integrating a diversity of crops and plants. Biodiversity improves the resilience of agricultural ecosystems.

Pollution reduction: By reducing the use of chemical fertilizers and pesticides, conservation agriculture minimizes soil and water pollution. Buffer zones and riparian buffer strips filter pollutants and protect watercourses.

Successful Case Studies

To illustrate the successes of conservation agriculture, let's look at a few notable case studies:

Direct seeding project in Brazil: In Brazil, the direct seeding project was set up to combat soil erosion and improve agricultural productivity. By adopting direct seeding and cover crops, farmers have succeeded in reducing erosion, increasing soil organic matter, and improving crop yields.

Soil Conservation Program in Kenya: In Kenya, a soil conservation program has been launched to help farmers adopt sustainable practices. Terracing, mulching and crop rotation techniques were introduced. Results showed a significant improvement in soil fertility, a reduction in erosion, and an increase in crop yields.

Agroforestry initiative in India: In India, an agroforestry initiative has been set up to integrate trees into farming systems. Farmers planted fruit trees and nitrogen-fixing trees among their food crops. The trees have improved soil fertility, provided additional food products, and increased farmers' incomes.

Water Management Program in Australia: In Australia, a water management program has been developed to improve irrigation efficiency and conserve water resources. Drip irrigation and rainwater harvesting techniques were adopted. Farmers succeeded in reducing their water consumption while maintaining high yields.

These examples show how targeted initiatives and sustainable practices can improve soil and natural resource conservation, while increasing agricultural productivity and protecting the environment. By adopting conservation agriculture, farmers can contribute to a more resilient, productive and sustainable agricultural future.

Chapter 27: Greenhouse farming

Greenhouse agriculture is an agricultural production method that enables plants to be grown in a controlled environment. This technique offers significant advantages for both intensive production and off-season cultivation, enabling farmers to increase yields, improve product quality and reduce the risks associated with adverse climatic conditions. This chapter explores the practices and technologies of greenhouse agriculture, and the benefits it offers for agricultural production.

Greenhouse farming practices

Greenhouse agriculture uses a variety of practices to create an optimal environment for plant growth. Here are some of the main practices:

Climate management: One of the main advantages of greenhouses is the ability to control the indoor climate. Growers can regulate temperature, humidity, ventilation and light to meet specific crop needs. Heating, cooling, shading and dehumidification systems are used to maintain ideal conditions.

Irrigation and fertilization: Greenhouse irrigation is often carried out using drip irrigation or micro-sprinkler systems, which provide precise, uniform water distribution. Fertigation, a technique that combines irrigation and fertilization, is commonly used to deliver nutrients directly to plant roots.

Pest and disease protection: Greenhouses offer physical protection against pests and diseases. Nets, screens and airtight doors keep pests out. Integrated pest management techniques, such as the use of natural predators and biological products, are also employed to minimize the use of pesticides.

Soilless cultivation: Soilless cultivation, or hydroponics, is a common greenhouse cultivation method. Plants are grown in inert substrates, such as rockwool, perlite or vermiculite, and receive a nutrient solution directly at their roots. This method allows precise control of nutrients and efficient use of water.

Automation and Technology: Modern greenhouses are often equipped with automation and control technologies. Sensors, climate management systems and monitoring software enable growers to monitor environmental conditions in real time and adjust parameters to optimize plant growth.

Greenhouse Agriculture Technologies

Greenhouse agriculture benefits from many advanced technologies that improve efficiency and productivity. Here are some of the main technologies used:

Climate Control Systems: Automated climate control systems regulate temperature, humidity, ventilation and lighting in greenhouses. These systems use sensors to monitor environmental conditions and software to adjust parameters in real time.

LED lighting: LED lighting is increasingly used in greenhouses to provide extra light for plants. LEDs are energy-efficient, long-lasting and can be adjusted to provide specific light spectra that promote photosynthesis and plant growth.

Drip Irrigation Systems: Drip irrigation systems enable precise distribution of water directly to plant roots. These systems reduce water wastage and improve irrigation efficiency. Fertigation can also be integrated into these systems to supply nutrients at the same time as water.

Hydroponics and aeroponics: Hydroponics and aeroponics systems are commonly used in greenhouses for soilless cultivation. In hydroponics, plants are grown in a nutrient solution, while in aeroponics, plant roots are suspended in the air and sprayed with a nutrient solution. These systems enable rapid growth and efficient use of resources.

Crop Management Systems: Crop management software helps growers plan, monitor and optimize their greenhouse operations. These systems provide data on plant growth, nutrient requirements, environmental conditions and crop performance, enabling informed decision-making.

Energy Management Systems: Modern greenhouses incorporate energy management systems to optimize energy use and reduce costs. Solar panels, efficient heating and cooling systems, and energy storage technologies are used to improve the energy sustainability of greenhouses.

Advantages of Greenhouse Farming

Greenhouse agriculture offers many advantages for agricultural production, particularly in terms of intensive and off-season production:

Intensified production: Greenhouses enable intensified agricultural production by providing optimal growing conditions. Farmers can grow plants at high densities, maximizing space utilization and increasing yields.

Out-of-season cultivation: One of the main advantages of greenhouses is the ability to grow crops out of season. By controlling the indoor climate, growers can produce crops all year round, regardless of the weather conditions outside. This makes it possible to meet market demand and generate stable income throughout the year.

Improved product quality: Controlled greenhouse conditions produce high-quality plants with uniform characteristics. Greenhouse-grown fruit, vegetables and flowers are often healthier, more attractive and have a longer shelf life.

Climate Risk Reduction: Greenhouses protect crops from extreme weather conditions, such as storms, frosts, droughts and floods. This protection reduces the risk of crop failure and improves food security.

Resource efficiency: Greenhouses enable more efficient use of resources such as water, nutrients and energy. Precise irrigation systems, fertigation technologies and LED lighting reduce waste and increase efficiency.

Pest and disease control: Greenhouses offer physical protection against pests and diseases. Integrated pest management techniques, combined with appropriate management practices, reduce the need for chemical pesticides and promote more sustainable crop protection methods.

Innovation and research: Greenhouses provide an ideal environment for innovation and research in agriculture. Farmers can experiment with new plant varieties, test advanced cultivation techniques and develop optimal management practices to improve productivity and sustainability.

Successful Case Studies

To illustrate the successes of greenhouse agriculture, let's look at a few notable case studies:

Les Serres de l'Almería in Spain: Spain's Almería region is famous for its vast greenhouses, which produce a large proportion of the vegetables consumed in Europe. The Almería greenhouses use advanced climate control, irrigation and soilless cultivation technologies to produce tomatoes, peppers, cucumbers and other high-quality vegetables. These greenhouses have transformed an arid region into a thriving agricultural production center.

Ontario Greenhouses in Canada: In Ontario, greenhouses are used to produce fresh vegetables, fruit and flowers all year round. Ontario greenhouses use efficient heating systems, LED lighting and crop management technologies to optimize production. Farmers benefit from high yields, high-quality produce and stable incomes.

Vertical Harvest greenhouses in the USA: Vertical Harvest is a US-based company that uses vertical greenhouses to produce fresh vegetables in urban environments. Vertical greenhouses maximize the use of space by growing plants at height. They use hydroponic technologies and climate control systems to produce vegetables all year round. Vertical Harvest contributes to urban food security and environmental sustainability.

Masdar City greenhouses in the United Arab Emirates: Masdar City, a sustainable city in the United Arab Emirates, uses greenhouses to produce fresh vegetables in a desert environment. Masdar City's greenhouses use passive cooling technologies, water management systems and solar panels to minimize the ecological footprint. These greenhouses show how innovation can overcome extreme climatic challenges and promote sustainability.

Practical Tips for Successful Greenhouse Farming

To succeed in greenhouse farming, it's essential to follow certain good practices and focus on key aspects. Here are a few practical tips:

Planning and design: Proper planning and design of greenhouses are crucial to their success. Choose a suitable location, select quality building materials and design effective ventilation and climate control systems.

Climate management: Regularly monitor and adjust climatic parameters such as temperature, humidity and light to create optimal growing conditions. Use automated systems to facilitate climate management.

Crop selection: Choose crops suited to greenhouse production and your specific climatic conditions. High-value crops such as tomatoes, peppers, cucumbers and herbs are often profitable in the greenhouse.

Irrigation and fertilization management: Use precise irrigation systems and fertigation techniques to supply plants with the water and nutrients they need. Monitor nutrient levels regularly and adjust applications accordingly.

Crop Protection: Implement integrated pest management strategies to protect crops against pests and diseases. Use nets, screens and natural predators to minimize the use of chemical pesticides.

Monitoring and tracking: Use sensors and management software to monitor environmental conditions, plant growth and crop performance in real time. Analyze data to make informed decisions and optimize operations.

Innovation and Training: Keep up to date with the latest innovations and technologies in greenhouse agriculture. Take part in training courses, workshops and conferences to improve your skills and knowledge.

By adopting these practices and using available technologies, farmers can maximize the benefits of greenhouse agriculture, improve production and contribute to the sustainability of agriculture. Greenhouse agriculture represents a promising solution for meeting global food challenges and ensuring a stable, high-quality food supply throughout the year.

Chapter 28: Types of farming

There are many different types of agriculture, each with its own techniques, objectives and advantages. Understanding the different types of farming is essential for farmers wishing to maximize their crop production, process their products efficiently and optimize their sales channels. This chapter explores the different types of farming, methods for maximizing agricultural crops, techniques for processing agricultural products, and sales channels available to farmers.

Types of farming

Agriculture can be classified into several types according to the techniques used, the scale of production, and the economic and ecological objectives. Here are some of the main types of agriculture:

Traditional agriculture: Traditional agriculture is based on ancestral techniques handed down from generation to generation. It uses mainly hand tools and natural methods to grow crops and raise livestock. This form of agriculture is often found in rural areas of developing countries.

Intensive agriculture: Intensive agriculture uses modern techniques and chemical inputs to maximize production on small surfaces. It relies on the use of machinery, fertilizers, pesticides and high-yielding plant varieties. This method is common in industrialized countries and aims to increase yields to meet growing food demand.

Organic farming: Organic farming avoids the use of synthetic chemicals and favours natural methods of crop and soil management. It uses organic fertilizers, crop rotations and biological control techniques to maintain healthy soils and ecosystems.

Precision Agriculture: Precision agriculture uses advanced technologies such as GPS, sensors and drones to optimize crop and resource management. It relies on the collection and analysis of data to make informed decisions and improve the efficiency of farming operations.

Agroforestry: Agroforestry combines agriculture and forestry by integrating trees and shrubs into crop and livestock systems. This approach promotes biodiversity, improves soil fertility and increases the resilience of agricultural systems.

Urban agriculture: Urban agriculture develops in urban and peri-urban areas, using techniques such as rooftop gardens, urban greenhouses and hydroponic systems. It aims to provide fresh, local food for urban communities.

Greenhouse agriculture: Greenhouse agriculture uses closed structures to create controlled environments, enabling plants to be grown all year round and outside the usual seasons. Greenhouses provide optimal climatic conditions for plant growth, increasing yields and product quality.

Subsistence farming: Subsistence farming is practiced mainly for the personal consumption of farmers and their families. It uses simple techniques and local resources to produce staple foods, ensuring food security for rural families.

Conservation Agriculture: Conservation agriculture focuses on preserving natural resources, particularly soil and water. It uses practices such as no-till, cover crops and crop rotation to improve the sustainability of farming systems.

How to maximize your crop

Maximizing crop production requires a combination of good farming practices, efficient resource management and advanced technologies. Here are some strategies for maximizing crop production:

Variety selection: Choosing plant varieties adapted to local conditions and production objectives is essential. Varieties resistant to disease, pests and extreme weather conditions can improve yields and crop resilience.

Soil management: Soil health is crucial to agricultural production. Using techniques such as adding organic matter, crop rotation and using organic fertilizers can improve soil fertility and increase yields.

Efficient irrigation: Irrigation is essential for maximizing yields, especially in arid regions. Drip irrigation systems and water management techniques can improve irrigation efficiency and reduce water losses.

Pest and Disease Control: Integrated Pest Management (IPM) combines biological, cultural and chemical methods to sustainably control pests and diseases. Using natural predators, resistant crops and appropriate management practices can minimize crop losses.

Precision Technologies: Using precision technologies, such as soil sensors, drones and crop management systems, enables accurate data to be collected and informed decisions to be made. This can improve the efficiency of farming operations and maximize yields.

Crop diversification: Diversifying crops can reduce the risk of crop failure and improve the resilience of farming systems. Crop rotations, intercropping and agroforestry are practices that promote biodiversity and sustainability.

Processing of agricultural products

Processing agricultural products adds value to raw materials, improves their shelf life and opens up new markets. Here are some common processing techniques:

Food processing: Food processing involves the conversion of agricultural raw materials into ready-to-eat food products. This includes processes such as milling cereals, making dairy products, preserving fruit and vegetables, and processing meat and fish. Food processing improves the nutritional value of products and reduces post-harvest losses.

Oil and fat production: Oilseeds such as soy, sunflower and rapeseed can be processed into vegetable oils. Oils can be used for cooking, in the manufacture of food products and in the cosmetics and pharmaceutical industries. The production of oils and fats is a lucrative business that adds value to raw materials.

Conversion to biofuels: Agricultural raw materials can be transformed into biofuels, such as biodiesel and ethanol. These biofuels are renewable alternatives to fossil fuels and can help reduce greenhouse gas emissions. Biofuel production can diversify farmers' sources of income.

Processing into cosmetics and pharmaceuticals: Many plants have medicinal and cosmetic properties. Agricultural raw materials can be transformed into pharmaceutical products, such as medicinal plant extracts, and cosmetics, such as creams, lotions and essential oils. Processing into cosmetics and pharmaceutical products is a high value-added activity.

Animal feed processing: Crop residues and agricultural by-products can be processed into animal feed. This processing adds value to by-products and provides a source of nutrition for livestock, contributing to the sustainability of farming systems.

Transformation into handicrafts: Agricultural raw materials can be transformed into handicrafts such as textiles, wooden objects and basketry. Transformation into handicrafts enables local resources to be valorized and traditional know-how to be preserved.

Sales channels

Sales channels play a crucial role in marketing agricultural products. Here are some common sales channels for farmers:

Local markets: Local markets are direct sales outlets where farmers can sell their produce directly to consumers. Local markets offer sales opportunities without intermediaries, enabling farmers to earn a larger share of their income. They also encourage interaction between producers and consumers.

Cooperatives and Producer Associations: Cooperatives and producer associations bring farmers together to market their products collectively. These organizations enable farmers to pool resources, negotiate better prices and access larger markets. They also offer training, credit and technical support services.

Supermarkets and supermarkets: Supermarkets and supermarkets are important sales channels for agricultural produce. Farmers can partner with these retailers to supply fresh and processed produce. Supermarkets offer wide consumer exposure and high sales volumes.

Online sales: Online sales are a fast-growing sales channel for agricultural products. E-commerce platforms enable farmers to sell their products directly to consumers via the Internet. Online sales offer extended geographical reach and the possibility of marketing niche products.

Supply contracts: Supply contracts are agreements between farmers and processors, retailers or exporters for the regular supply of agricultural produce. These contracts offer income security to farmers and guarantee a constant supply to buyers.

Short circuits and buying groups: Short circuits, such as vegetable baskets and community buying groups, enable farmers to sell their produce directly to local consumers. These initiatives encourage consumption of local produce, cut out the middleman and strengthen links between producers and consumers.

Exporting: Exporting is a sales channel for agricultural products destined for international markets. Farmers can export fresh, processed or artisanal products to foreign countries. Exporting offers opportunities for growth and income diversification, but requires compliance with international quality and food safety standards.

Strategies for Optimizing Sales Channels

To maximize income and reach a wide range of consumers, farmers can adopt the following strategies:

Diversifying sales channels: Diversifying sales channels helps to reduce risk and reach different market segments. Farmers can combine direct sales, cooperatives, supermarkets and online sales to optimize their income.

Improving Quality and Traceability: Providing high-quality products and ensuring product traceability are key factors in winning the confidence of consumers and buyers. Farmers need to adopt quality management and certification practices to meet market expectations.

Marketing and Promotion: Investing in marketing and promotion strategies helps increase product visibility and attract consumers. Farmers can use digital marketing tools, advertising campaigns and promotional events to promote their products.

Innovation and Differentiation: Innovating and differentiating products helps them stand out in the marketplace. Farmers can develop niche products, unique varieties or processed products to meet consumer trends and preferences.

Training and skills development: Ongoing training and skills development are essential for success in the marketing of agricultural produce. Farmers need to keep abreast of market developments, technologies and management practices to improve their performance.

Collaboration and Partnerships: Collaborating with other farmers, businesses, research institutions and support organizations can strengthen marketing capacity. Partnerships help pool resources, share knowledge and develop innovative solutions.

By adopting these strategies and exploiting the various sales channels available, farmers can maximize the value of their products, reach a wide range of consumers and improve their profitability. Modern agriculture offers many opportunities to innovate, diversify and optimize farming operations, thus contributing to food security and the sustainability of farming systems.

Chapter 29: Conclusion

Agriculture is a vital sector for food security, economic development and environmental sustainability. Through this guide, we have explored a variety of farming practices, product processing techniques and marketing strategies. This conclusion synthesizes the main points discussed and offers thoughts on the future of agricultural processing and the types of farming for sustainable, profitable production.

Summary of Main Points Discussed

Diverse types of agriculture: We've looked at different types of agriculture, each with its own techniques, advantages and challenges. Traditional agriculture relies on ancestral methods, while intensive farming uses modern technologies to maximize yields. Organic farming avoids synthetic chemicals, precision farming uses advanced technologies to optimize crop management, and agroforestry integrates trees into farming systems to enhance biodiversity and soil fertility. Urban and greenhouse agriculture offer solutions for urban and off-season production, while subsistence and conservation agriculture focus on food security and the sustainability of natural resources.

Maximizing Crop Production: Maximizing crop production requires a combination of good farming practices, efficient resource management and advanced technologies. The selection of adapted varieties, soil management, efficient irrigation, pest and disease control, and the use of precision technologies are key strategies for improving crop yields and resilience.

Processing agricultural products: Processing agricultural products adds value to raw materials, improves their shelf life and opens up new markets. Food processing techniques, the production of oils and fats, biofuels, cosmetics and pharmaceuticals, animal feed and handicrafts are all ways of diversifying income sources and improving farm profitability.

Sales channels : Sales channels play a crucial role in the marketing of agricultural products. Local markets, cooperatives, supermarkets, online sales, supply contracts, short circuits and exports offer a variety of opportunities for farmers to sell their produce. Diversifying sales channels, improving quality and traceability, investing in marketing and innovation, and collaborating with other players in the sector are key strategies for optimizing income.

Reflections on the Future of Agricultural Transformation and Types of Agriculture

The future of agriculture lies in farmers' ability to innovate, adopt sustainable practices and respond to global challenges such as climate change, population growth and food security. Here are some thoughts on the future of agricultural transformation and types of farming for sustainable, profitable production:

Adoption of Advanced Technologies: Advanced technologies such as precision farming, automation, sensors and drones will continue to play a central role in improving farm efficiency and productivity. The adoption of these technologies will enable farmers to make decisions based on accurate data, optimize the use of resources and reduce environmental impacts.

Sustainable farming practices: Sustainability will be a priority for the future of agriculture. Sustainable agricultural practices, such as organic farming, agroforestry, conservation agriculture and regenerative agriculture, will be essential to preserve natural resources, improve soil health and promote biodiversity. Supportive policies and economic incentives will encourage the adoption of these practices.

Food processing and nutritional security: Food processing will play a crucial role in improving nutritional security and reducing post-harvest losses. Innovations in processing techniques, food preservation and the production of high value-added products will help meet the growing demand for healthy, nutritious food.

Marketing and market access: Market access will be essential for farm profitability. Farmers will need to diversify their sales channels, improve product quality and traceability, and invest in marketing and promotion strategies. E-commerce platforms and short circuit initiatives will offer opportunities to reach a wide range of consumers.

Resilience and Adaptation to Climate Change: Agriculture will need to adapt to the impacts of climate change, such as temperature variations, erratic rainfall and extreme weather events. Resilient agricultural practices, resistant plant varieties and climate risk management systems will be essential to maintain agricultural production and ensure food security.

Collaboration and Partnerships: Collaboration between farmers, researchers, support institutions and businesses will be crucial to the development of innovative and sustainable solutions. Public-private partnerships, collaborative research initiatives and knowledge-sharing networks will foster the adoption of advanced farming practices and the resilience of the agricultural sector.

Investment and support policies: Investment in agricultural infrastructure, research and development, and training and support programs will be essential for the future of agriculture. Supportive policies, such as subsidies, tax incentives and funding programs, will encourage innovation and the adoption of sustainable practices.

Case Studies and Inspiring Examples

To illustrate these thoughts, let's take a look at some inspiring case studies and examples:

Vertical farms in Asia: In Asia, vertical farms have become an innovative solution for urban food production. Using hydroponic technologies and advanced climate control systems, these farms produce fresh vegetables all year round, reducing dependence on food imports and improving urban food security.

Soil Regeneration Initiative in Africa: In Africa, a soil regeneration initiative has been launched to restore degraded land and improve agricultural productivity. Farmers are adopting practices such as agroforestry, direct seeding and composting to enrich soils and boost yields. This initiative contributes to food security and the resilience of rural communities.

Brazil's Biofuels Program: Brazil is a world leader in the production of biofuels from sugarcane and soybeans. Brazil's biofuels program has diversified farmers' sources of income, reduced dependence on fossil fuels and contributed to the reduction of greenhouse gas emissions. This program shows how the processing of agricultural products can support sustainable development objectives.

Agricultural cooperatives in Europe: In Europe, agricultural cooperatives play a crucial role in the marketing of agricultural products. These cooperatives bring farmers together to pool resources, negotiate better prices and access larger markets. They also offer training, credit and technical support services, strengthening farmers' ability to innovate and develop.

Food security programs in India: In India, food security programs have been set up to improve the nutrition and resilience of rural families. These programs include vegetable gardening, crop diversification and food processing initiatives. They aim to reduce malnutrition, increase farmers' incomes and promote sustainable agricultural practices.

These examples show how innovation, collaboration and sustainable practices can transform agriculture and meet global challenges. By adopting advanced strategies and exploiting available opportunities, farmers can contribute to a more resilient, productive and sustainable agricultural future.

Final conclusion

Agriculture is a constantly evolving sector, facing complex challenges and promising opportunities. By understanding the different types of agriculture, adopting sustainable practices, harnessing advanced technologies and optimizing sales channels, farmers can maximize production, improve profitability and contribute to global food security. This practical farming guide aims to provide advice, strategies and inspiring examples to help farmers navigate this ever-changing landscape and succeed in their efforts towards sustainable, profitable agriculture.

Chapter 29: Conclusion

Throughout the chapters of this practical farming guide, we have explored a wide range of methods, technologies and strategies designed to maximize agricultural production, process produce in innovative ways and optimize sales channels. The diversity of farming practices presented here, from traditional to precision farming, organic to greenhouse, underlines the importance of adaptability and innovation in meeting contemporary challenges in the agricultural sector.

The future of agriculture lies in farmers' ability to innovate, adopt sustainable practices and respond to global challenges such as climate change, population growth and food security. Precision agriculture, with its advanced technologies such as sensors, drones and crop management systems, represents a promising avenue for optimizing the use of resources and improving productivity. The adoption of these technologies will enable farmers to make decisions based on accurate data, optimize the use of resources and reduce environmental impacts.

At the same time, sustainable agricultural practices will become increasingly essential. Organic farming, agroforestry, conservation agriculture and regenerative agriculture are not only viable alternatives, but necessities for preserving natural resources, improving soil health and promoting biodiversity. Supportive policies and economic incentives will play a crucial role in encouraging the adoption of these practices. Processing agricultural products adds value to raw materials, extends their shelf life and opens up new markets. Food processing techniques, the production of oils and fats, biofuels, cosmetics and pharmaceuticals, as well as animal feed, are all ways of diversifying income sources and improving farm profitability.

Access to markets is another crucial aspect of farm profitability. Local markets, cooperatives, supermarkets, online sales, supply contracts, short circuits and exports offer a variety of opportunities for farmers to sell their produce. Diversifying sales channels, improving quality and traceability, investing in marketing and innovation, and collaborating with other players in the sector are key strategies for optimizing income. In this context, resilience and adaptation to climate change must be at the heart of agricultural concerns. Resilient farming practices, resistant plant varieties and climate risk management systems will be essential to maintain agricultural production and ensure food security. Farmers will need to adapt to the impacts of climate change, such as temperature variations, irregular rainfall and extreme weather events.

Collaboration between farmers, researchers, support institutions and businesses will be crucial to the development of innovative and sustainable solutions. Public-private partnerships, collaborative research initiatives and knowledge-sharing networks will foster the adoption of advanced farming practices and the resilience of the agricultural sector. Finally, investment in agricultural infrastructure, research and development, and training and support programs will be essential for the future of agriculture. Supportive policies, such as subsidies, tax incentives and financing programs, will encourage innovation and the adoption of sustainable practices.

Vertical farms in Asia, soil regeneration initiatives in Africa, the biofuels program in Brazil, agricultural cooperatives in Europe and food security programs in India are inspiring examples of what innovation and collaboration can achieve. These examples show how sustainable practices, advanced technologies and effective marketing strategies can transform agriculture, improve food security and promote sustainability.

In conclusion, agriculture is a constantly evolving sector, facing complex challenges and promising opportunities. By understanding the different types of agriculture, adopting sustainable practices, harnessing advanced technologies and optimizing sales channels, farmers can maximize production, improve profitability and contribute to global food security. This practical farming guide aims to provide advice, strategies and inspiring examples to help farmers navigate this ever-changing landscape and succeed in their efforts towards sustainable, profitable agriculture.

References

A- Food Processing Technology: Principles and Practice by P.J. Fellows

This book covers the basic principles and techniques of food processing, including dairy products, grains, and more. "Dairy Science and Technology" by P. Walstra

An in-depth resource on dairy processing methods and technologies. "Handbook of Cereal Science and Technology" by Karel Kulp and Joseph G. Ponte Jr.

A comprehensive guide to cereal processing. "Sugarcane: Agricultural Production, Bioenergy, and Ethanol" by Fernando Santos and Luiz Fernando Cortez.

This book covers sugar crop processing and economic impact. "Spices, Herbs and Edible Fungi" by G. Charalambous

Spice and herb processing techniques. "Edible Oils and Fats: Developments Since 1986" by R.J. Hamilton

A guide to oilseed processing. "
Beer: Tap into the Art and Science of Brewing" by Charles Bamforth

A detailed introduction to brewing. "Understanding Wine Technology" by David Bird

A guide to winemaking.

B- Transformation into Chemical Products
"Biomass to Biofuels: Strategies for Global Industries" by Alain A. Vertes et al.

The process of producing biofuels from agricultural materials. "Cosmetic Science and Technology: Theoretical Principles and Applications" by Kazutami Sakamoto et al.

Transformation of agricultural materials into cosmetic products. "Pharmaceutical Biotechnology: Fundamentals and Applications" by Daan J.A. Crommelin et al.

Transformation of agricultural products into pharmaceutical ingredients. "Biofuels: Securing the Planet's Future Energy Needs" by Adrian Dumitru-Tanase

Production of biodiesel and bioethanol.
"Starch: Chemistry and Technology" by James BeMiller and Roy L. Whistler

Starch processing methods.
"Enzyme Technology" by Martin F. Chaplin and Christopher Bucke

Use of enzymes for chemical transformation. "Biopolymers: Biology, Chemistry, Biotechnology, Applications" by Alexander Steinbüchel.

Polymerization processes of agricultural materials. "
The Biodiesel Handbook" by Gerhard Knothe et al.
Transesterification
techniques for biodiesel production.
"Green Chemistry: Theory and Practice" by Paul T. Anastas and John C. Warner

Principles of green chemistry applied to agriculture.
"Natural Fibers, Biopolymers, and Biocomposites" by Amar K. Mohanty et al.

Production of natural fibers from agricultural materials.

C- Different Types of Agriculture
"Traditional Agriculture: A Critical Examination" by P.A. McAnany and B.L. Turner II

Traditional farming methods. "
Intensive Agriculture and Sustainability: A Farming Systems Analysis" by Glen C. Filson and B. Adekunle

Practices and impacts of intensive agriculture. "
The Organic Farming Manual" by Ann Larkin Hansen

Techniques and principles of organic farming.
"Urban Agriculture: Ideas and Designs for the New Food Revolution" by David Tracey

Development of urban agriculture. "Agroforestry for Sustainable

Agriculture" by Miguel A. Altieri et al.

Benefits of agroforestry. "Precision Agriculture Technology for Crop Farming" by Qin Zhang

Use of advanced technologies in agriculture.
"Subsistence Agriculture and Economic Development" by Clifton R. Wharton Jr.

Subsistence farming practices. "
Conservation Agriculture: Global Prospects and Challenges" by Muhammad Farooq et al.

Soil and natural resource conservation methods.
"Greenhouse Gardening: Step by Step to Growing Success" by Alan Titchmarsh

Table of Contents

MIX
Papier aus verantwortungsvollen Quellen
Paper from responsible sources
FSC® C105338

Printed by Books on Demand GmbH, Norderstedt / Germany